SHOULD WE RISK IT?

Should We Risk It?

Exploring Environmental, Health, and Technological Problem Solving

Daniel M. Kammen and
David M. Hassenzahl

PRINCETON UNIVERSITY PRESS

PRINCETON, NEW JERSEY

ISBN 0-691-00426-9

Library of Congress Cataloging-in-Publication Data

Kammen, Daniel M., 1962
    Should we risk it? : exploring environmental, health, and technological
      problem solving / Daniel M. Kammen and David M. Hassenzahl.
      p.   cm.
    Includes bibliographical references and index.
    ISBN 0-691-00426-9 (cloth : alk.paper)
    1. Environmental sciences—Decision making. 2. Environmental
      policy—Government policy. 3. Risk assessment. 4. Health risk
      assessment. 5. Technology—Risk assessment. 6. Science and state.
    I. Hassenzahl, David M. II. Title.
GE105.K35 1999
363.7′02—dc21   98-49393

Dedicated to

Dele and Sadé
and
Hilary and Mikaela

Contents

Preface

Motivation for This Book

Demand for risk analysts, especially in technological arenas such as environmental and health sciences, has increased exponentially in the past several years. The range of fields where risk analysis provides useful and relevant tools has expanded to the point that risk modeling and forecasting is now a basic operational tool across the physical, biological, social, environmental, and health sciences. Unfortunately, programs that train these analysts have not kept pace with this demand. Consequently, many people who find themselves in positions where they must informally make decisions about risk also find that they must learn the tools of that trade on the fly. While it is always true that practice differs from theory, better training in the fundamentals of problem solving and risk analysis can be easily achieved in one or two semesters of an undergraduate or professional graduate program, or alternatively in a supplementary semester course for those Ph.D. students in traditional disciplines ranging from physics to psychology who expect to be engaged in risk analysis.

Like any discipline, risk analysis taught in lecture form can be highly formal, or worse, very dry material. Similarly, neither discussion formats nor lectures prepare the student for the main "real-world" application—problem solving. When risk analysis problems are available they are often unsolved, lack the data necessary to solve them, and/or are not real-world problems. By contrast, this book provides complete solutions for representative problems, including the data needed to answer all sections. Many of the problems are taken directly from real-world events, and the rest are realistic situations designed to demonstrate some important facets of risk analysis. A central motivation for this book is the experience that each of us has had in learning to solve environmental problems.

John Harte, a professor at the University of California at Berkeley's Energy and Resources group, faced this same problem in teaching environmental problem solving. His solution, the book *Consider a Spherical Cow: A Course in Environmental Problem Solving*, is the inspiration and in many ways the model for this book. Harte's book, known affectionately to those who have learned or taught from it as *Cow*, has now shown students for a decade not merely how to crunch numbers, but how to approach diverse problems in effective and efficient ways. At the same time, *Cow* illustrates a method to teach these tools in a much more stimulating fashion than lectures and standard textbooks can provide. Our overly ambitious goal is to emulate the spirit of *Cow* for risk analysis.

The authors also have personal motivations for having a book like this available. Hassenzahl earned his undergraduate degree at the University of California at Berkeley, firmly grounded in engineering, physical, and biological science as well as social science. His education included learning the tools in *Cow* in a class co-taught by John Harte and John Holdren, now in the Center for Sciences and International Affairs at Harvard's Kennedy School of Government. While the program provided him with a better foundation in risk analysis than most, he nonetheless had to learn much on his own in his subsequent jobs.

In his first position, as an environmental specialist at an industrial facility, his responsibilities included air, water, and solid waste risk management. He had to teach himself how risk analysis was done in order to ensure regulatory compliance in day-to-day operations, oversee mandated risk analyses, and negotiate operating permits. In his second job at a regulatory agency, his role included ensuring regulatory compliance at a range of facilities. As such, he was often a professional risk communicator, maintaining dialogue between industry and the public, yet never received any formal training in that area.

For Kammen, this book has both immediate and long-standing origins. Trained formally as a physicist, Kammen was introduced to, and then began learning, risk assessment in the chaotic style of a disciplinary wanderer. While still a doctoral student in physics, his interests shifted to energy, environment, and development topics. Armed with few tools to explore these topics except a physicist's inclination to build models (and then not to trust them beyond the empirical data), he found that risk analysis provided a natural first window on issues that clearly required a richer analysis. He was tremendously fortunate to receive encouragement and guidance from a number of people, in particular José Goldemberg, John Graham, John Harte, John Holdren,

Kirk Smith, Rob Socolow, and Richard Wilson, who took the time to help a novice informally explore topics in risk as they applied to the specific energy and environment questions he was studying at the time. The first questions to which he applied these tools were the health and environmental impacts of new, renewable energy in solar thermal cooking and heating technologies, the second the "dose-response" relationship between global warming and sea-level rise. There could be no better, or less structured, education.

Two years later, while a postdoctoral fellow at the California Institute of Technology, Kammen gave a talk on health risks from solar ovens and other improved cookstove technologies. Following this talk, one student asked him whether he, as a physicist, studied risk analysis. Kammen's response was the meandering paragraph above; that he studied it piecemeal and informally as a graduate student and postdoctoral fellow. The student's reaction, one of perplexed interest, expressed the question, "But where do *I* study risk?" The perception of a lack of practical courses is quite common.

An example of the type of training that is becoming available is at Columbia University's School of Public Health. When Dr. Joseph Graziano became head of that school's Division of Environmental Health Sciences in 1991, his interviews with students and alumni led to the addition of risk assessment to the curriculum. While a course has been established, Dr. Graziano notes that there is no standard textbook and a paucity of real-world problems to stimulate the students. Many other schools are in the same situation: they have begun teaching classes on risk analysis, but find a lack of good teaching materials. Instructors who wish to assign problem sets must create them, a quite time-consuming project.

At a more immediate and specific level, the need for this book emerged from the "Methods in Science and Technology Policy" course that Kammen offers at Princeton University. This course includes students of public policy from a range of backgrounds: scientists, engineers, and social scientists. The challenge is to teach the techniques of risk assessment in a form practical enough to benefit this diverse audience. For Kammen, a set of real-world problems offers the clearest means to illustrate the range of techniques necessary to make headway on interdisciplinary, and often ill-posed or incomplete, problems. A number of the problems collected here stem from the problem sets and in-class exercises from four years of offering this course, and a draft of this book has been used for the same class taught by visiting faculty member Tony Nero.

Once this project focused into an effort to produce a book, a wider range of risk assessment teachers came into play. In an effort to increase the pool of worked problems and the range of problem-solving styles, the authors discussed the project with colleagues at many institutions and organizations, including Princeton University, the University of California at Berkeley, Harvard University, the Massachusetts Institute of Technology, Clark University, the California Institute of Technology, Cornell University, the University of California at Los Angeles, the University of Pennsylvania, Carnegie-Mellon University, Resources for the Future, the Society for Risk Analysis, Lawrence Berkeley Laboratory, and the U.S. Environmental Protection Agency. Risk courses, or at least problems, were found lurking everywhere. Some of these problems have been gratefully borrowed and occasionally modified, and appear in the chapters that follow. To this group of contributors, the authors owe a huge and ongoing debt, one that is listed, albeit incompletely, in the acknowledgments section.

A book like this is never done. Alternative solutions and new problems are always appearing. In an effort not only to correct errata, but more importantly to gather and distribute new material, we ask the reader to send these to us. We will shamelessly use many of them (but with full attribution!) in future editions of this book.

Acknowledgments

This book is as much a testament to the many individuals who have encouraged and assisted us as to the diverse methods and ways of thinking about risks to our environment. This project is also, ominously, never done. Corrections and new routes to solving these problems are always welcome. New methods and techniques, and new case studies, appear all the time, particularly as risk assessment is used and adapted to an ever-wider range of disciplines and problem areas. We cannot hope to keep this volume up to date with all these changes, but with input in the form of feedback or new problems from other practitioners and students of risk, we are game to try.

There are a variety of intellectual debts we owe in this project, but they largely begin with John Harte who provided the model in *Consider a Spherical Cow*. We hope that this book is not too far off the mark he set in *Consider a Spherical Cow*, at least in spirit. Suggestions and problems were provided by many people, but a listing, sure to be incomplete, includes Clint Andrews, Tom Beierle, Claire Broido, Tony Cox, Weihsueh Chiu, Terry Davies, Adam Finkel, John Graham, Bill Hassenzahl, Sr., John Holdren, Henry Horn, Sheila Jasanoff, Danny Kahneman, Michael and Carol Kammen, George Kassinis, Willett Kempton, Vic Kimm, Ann Kinzig, Howard Kunreuther, Ed Lyman, Robert Margolis, Rachel Massey, David Romo Murillo, Bill Pease, Archis Parsharami, Paul Portney, Phil Price, Amy Richardson, Eldar Shafir, Doris Sloan, Rob Socolow, Valerie Thomas, Jim Wilson, Dick Wilson, and Wolf Yeigh.

Baruch Fischhoff, John Harte, James Hammitt, Kristin Shrader-Frechette, Tony Nero, Majid Ezzati, and David Hauri all commented at length on drafts. Their encouragement and expert advice led us to a broader, yet more purposeful, whole. We could not have found a better set of reviewers.

Trevor Lipscombe, our editor at Princeton University Press, has provided invaluable guidance, uncovered a wonderful French print for the book's cover, and gently prompted us to stay at least near our schedule.

Karen Verde at Princeton University Press tolerated tedious questions and last-minute changes, and Jennifer Slater's copy editing was extraordinary.

Dele and Sadé Kammen and Hilary and Mikaela Hassenzahl, the dedicatees of this volume, deserve deepest gratitude for their patience and support.

Hassenzahl thanks Clint Andrews and Jim Doig for their patience in overlooking other work that was *not* done during the writing of this book.

The students from four years of Kammen's course WWS-589, "Methods in Science, Technology and Public Policy" at Princeton were the unwitting guinea pigs (heterogeneous test animals?) for many of these problems and provided important feedback. Tony Nero and Rick Duke provided extensive and invaluable comments, while leading a fifth batch of WWS-589 students through a draft.

Majid Ezzati wrote several problems and solutions for WWS 589, wrote or helped with a number of others, and reviewed a number of draft chapters. Dan Koznieczny did a remarkable job of testing, building, and evaluating both problems and text, and Carter Ruml undertook a test run of the entire book. Weihsueh Chiu, Baruch Fischhoff, Rob Goble, Dale Hattis, Frank von Hippel, Tracy Holloway, and Tony Nero all provided extraordinary problems. Catherine Dent subjected herself to early chapters, substantially improving readability. Jackie Schatz has helped immeasurably with the entire writing process.

With so much help, any errors and omissions that remain are all the more embarrassing, and, of course, are ours alone.

SHOULD WE RISK IT?

1

Introduction

I began by trying to quantify technical risks, thinking
that if they were "put into perspective" through
comparison with familiar risks we could better judge
their social acceptability. I am ashamed
now of my naiveté, although I have the excuse
that this was more than twenty years ago,
  while some people are still doing it today.

                                      Harry Otway, 1992

Defining Risk

What is risk? What are the tools and methods used to evaluate
particular health, environmental, technological, and other risks, and
what are the limitations, uncertainties, and biases in these methods?
How can and will the results found using those methods be used by
individuals and groups?

This book is about modeling and calculating a variety of risks,
understanding what we're trying to calculate, and why we would want to
do so. First, however, what is risk? A simple, albeit "technocratic,"
definition of that risk is the probability that an outcome will occur times
the consequence, or level of impact, should that outcome occur. To
many people, risk suggests adverse outcomes; however, technical approaches to evaluating probabilities and outcomes are not limited to
negative impacts. Rather, they represent positive or negative changes in
state.

We can quantify risks in a number of ways, and often with considerable precision. While this quantification can be a useful tool, it is not
the whole story. This book leads through technical and analytic methods
used to evaluate and test risk, and then into the more intricate world of
social valuation and decision theory to which Otway alludes. We begin
with an exploration of the quantitative methods, and then expand the
sphere of analysis to include uncertainty, economic, political, and social
dimensions of risk understanding and management. Our operating

principle is that when we can better understand and describe values (that is, what the outcomes and probabilities are likely to be and how complete our understanding is), we can make better decisions.

Sheila Jasanoff proposes that the role of risk assessment is to "offer a principled way of organizing what we know about the world, particularly about its weak spots and creaky joints" (Jasanoff 1993). In keeping with this philosophy, the goal of this book is not to produce "technocrats" who will apply these tools to decisions outside of a social context. Rather, we hope that our readers will learn not only how to "crank the numbers," but when and why they should, and how the numbers will be interpreted in a broader cultural context. Ideally, risk analysis responds to the needs of interested and affected groups and individuals; it is intended to inform, but not determine, decisions.

Examples of the pressing need for better risk analysis abound. At the microdecision level, this agenda includes evaluating the impacts of and possible responses to rare but potentially "catastrophic" risks; identifying mechanisms of disease (and consequently improving opportunities to cure or avoid them); comparing similar remedies to a single adverse situation; and evaluating the possibly different responses of adults and children to a potential risk factor.

This book introduces a diverse audience to the fundamental theories and methods for modeling and analyzing risk. As a synthetic approach to both the subject of risk and the standard risk analysis "tool kit" we envision the potential for wide use in the fields of environmental science, engineering risk/fault analysis, public policy and management, and science policy. In particular, these methods should be of interest to policy makers at the local, state, and federal level who are now confronted with legislation that requires them to perform risk and cost/benefit analyses prior to a range of actions.

Increasingly, professional decision makers such as engineers, environmental scientists, "policy wonks," and others find that they need to answer risk questions. They may be asked to generate a report on risks, or to recreate and critique how someone else created a report. They may need to be able to communicate their work to a skeptical public, or to a busy politician. They are also likely to find that they lack the tools to deal with these issues as they arise.

At the same time, the uninitiated are likely to see the process of risk assessment as enormously complex and problem specific. Looking at a single problem too closely can lead to two unsatisfactory end points.

One is to leave the problems "up to the experts," taking the results from the risk assessment "black box" at face value. The other is to get lost in the details of the problem at hand. This is unfortunate, since a few general tools can equip analysts to tackle most, if not all, problems of risk.

The fields of science and technology policy and environmental studies have only a limited number of unifying methods. The goal of our work is to develop a practical approach to formulating, solving, and then generalizing the theory and methods of risk analysis. This book provides a set of tools to clarify and define these methods, producing more than the current set of fascinating, but idiosyncratic and anecdotal, case studies. We seek to bridge the gap between qualitative "discussion" books, which provide little analytic or practical training; advanced modeling books and journal papers, which generally assume considerable prior knowledge on the part of the reader; and highly specialized works in the areas of medical epidemiology or industrial emissions. To do so, we present and suggest solutions to real-world problems using a variety of risk analysis methods.

The case studies we present include subjects as diverse as the health impacts of radon, trends in commercial and military flight safety, extrapolation from high-dose laboratory animal studies to low-dose human exposures, and some key decisions relevant to the proposed national high-level nuclear waste storage facility at Yucca Mountain, Nevada. The solutions to the exercises provide a springboard to the broader applications of each method to other technological, environmental, public health, and safety risk issues, as well as to forecasting and uncertainty. Additional unsolved problems reinforce the presentation. The methods include the scientific and quantitative methods used to evaluate risks, as well as analytical tools for social/political management and decision making.

The central theory and methods of risk covered in this book include order-of-magnitude estimation; cause-effect (especially dose-response) calculations; exposure assessment; extrapolations between experimental data and conditions relevant to the case being addressed; modeling and its limitations; fault-tree analysis; and managing and estimating uncertainty. While not the central focus of this book, statistics play a key role as a basic tool. We cover basic and intermediate statistics in chapter 3. Probabilistic risk assessment (PRA) methods, Bayesian analysis, and various techniques of uncertainty and forecast evaluation are presented and used throughout the book.

Note that we do not address the expanding field of financial risk. While many of the models and techniques are similar to those presented here, there is an entire literature devoted to that subject.

Structure of the Book

The goal of this book is to introduce the student to advanced risk analysis tools, but we believe that the risk analyst must be able to walk before she can run. In other words, gaining proficiency in the fundamentals of risk analysis necessarily precedes deeper understanding, and even mastering the basics can substantially aid decision making. Consequently, most of the book is directed at learning to manipulate various individual tools, and understanding their applications and limitations. Toward the end of the book we provide examples of real-world applications ranging from local, specific, and clearly definable risks to some that involve multiple stakeholders and substantial uncertainty. The remainder of this introductory chapter discusses the history of the risk policy process, the current status of risk analysis as a central but often ad hoc technique, and the main areas of agreement and dispute about definitions and methods.

The first section (chapters 2–4) covers the basic "tools of the trade." Chapter 2 presents basic modeling techniques, both with and without numbers. The use of "stock and flow" models as an approach to identifying and quantifying exposures is presented first, followed by a number of models and techniques for quantifying cause-effect relationships.

Chapter 3 reviews the basic statistical techniques most commonly used in risk assessment. In general, solving the problems in this book requires fluency in high-school mathematics and basic statistics. For some problems calculus is a useful, although not necessary, prerequisite. (In fact, given the extent of uncertainty involved in many risk decisions, it should become clear that *over*-analysis can be a real problem.) While some of the models are easier to manipulate using more advanced mathematics, all the concepts and much of the implementation should be within the grasp of most college students. Many of the problems in this book have been used in the Princeton University graduate course

"Methods in Science and Technology Policy" (WWS-589), and have been taught without reference to calculus.

The beginning of the statistics chapter, designed more as a text than the rest of the book, is intended to be a review for those whose statistics are rusty; for the novice, a basic statistics class or text is recommended. The fourth chapter concludes the basic tools section with a discussion of variability, uncertainty, and forecasting, and provides two sophisticated statistical tools for dealing with variability and uncertainty: Bayesian analysis and probabilistic (Monte Carlo) analysis.

The second section of the book (chapters 5–8) applies these techniques to four important risk methodologies: structural models (e.g., toxicology), empirical models (e.g., epidemiology), exposure assessment, and technological risk assessment. Many of the problems address environmental risk, simply because that is where the authors have the most experience. However, a range of other issues are included, as well as discussion of how these methods can be applied in other fields.

The final section (chapters 9 and 10) deals with social aspects of risk: how people perceive risks, how people learn and communicate about risk, and how risk assessment can be incorporated into private and public decisions. The ninth chapter reiterates that the *application* of these tools should be limited to, motivated by, and designed to inform stakeholder and policy needs. This chapter puts the rest of the book into the decision-making context, introducing and critiquing some formal methods for both comparing among diverse risks and incorporating diverse interests. The final chapter discusses the human agent, and how perceptions of risk by both experts and nonexperts, as well as risk communication methods, influence risk decisions.

Risk analysis and computers complement one another very well, and most risk classes we are aware of incorporate a variety of software packages. Several of the problems in this book require the use of spreadsheets and risk software. In writing the problems, the authors generally used Microsoft Excel and the Crystal Ball and solver.xls add-ins, but other packages (such as Stella and @Risk) are of course acceptable.

Even small risk decisions may require many steps. No single problem or chapter can make the reader a "fully qualified risk analyst," but as a whole this book should enable the reader to synthesize the individual steps, combining them into coherent decisions. It will also promote enough healthy skepticism to guard against blind faith in any single methodology.

A book like this can never be truly "final." New solutions to old problems may be proposed by the readers, new information may change an existing problem, and emerging risks suggest novel methods and exercises. To keep pace, we are maintaining a website for this book at http://socrates.berkeley.edu/erg/swri. At the site, you will find

- Updated versions of problems and solutions in the book, including downloadable data files
- Copies of supplemental problems
- Solutions to supplemental problems, available to registered course instructors (if you are teaching from this book, contact the authors for a password)
- A dialog box to comment, append, or correct existing problems
- A dialog box to enter new problems and/or solutions

Our hope is that readers will contribute new cases that we will make available both on the World Wide Web site and in future editions of the book (with full attribution) as the fields of risk analysis and policy evolve.

Risk Analysis and Public Policy

In the past several decades, formal risk analysis has played an increasingly influential role in public policy, from the community to the international level. Although its outputs and uses are often (even usually) contentious, it has become a dominant tool for energy, environmental, health, and safety decisions, both public and private. More recently, risk analysis and cost-benefit analysis have been suggested by some (and even debated in Congress) as the *principal* tools for major federal environmental decisions, while others argue that the two methods have been oversold. While critiques abound, few scholars and practitioners would dispute the notion that an understanding of some essential tools of the trade is invaluable.

Risk analysis in one form or another has been used for centuries (see box 1-1, taken from Covello and Mumpower 1985). In the early 1970s, as risk analysis evolved into a major policy decision tool, Alvin Weinberg (1972) proposed that it falls into a special category of "trans-science . . . questions which can be asked of science, yet which cannot be

---

Box 1-1. Some historical highlights in risk analysis

About 3200 B.C.: The Asipu, a group of priests in the Tigris–Euphrates Valley establish a methodology:

- Hazard identification
- Generation of alternatives
- Data collection* and analysis
- Report creation

*Note that "data" included signs from the gods!

Arnobius, 4th century A.D., came up with decision analysis and first used the *dominance principle*, whereby a single option may be clearly superior to all others considered. Arnobius concluded that believing in God is a better choice than not believing, whether or not God actually exists. Note that Arnobius did not consider the possibility that a different God exists.

		State of nature	
		God exists	No God
	Believe	Good outcome (heaven)	Neutral outcome
Alternative			
	Don't believe	Bad outcome (hell)	Neutral outcome

King Edward II had to deal with the problem of smoke in London:
 1285: Established a commission to study the problem.
 1298: Commission called for voluntary reductions in use of soft coal.
 1307: Royal proclamation banned soft coal, followed by a second commission to study why the proclamation was not being followed.

---

answered by science." Individuals and society need to make decisions on issues for which there are no certain outcomes, only probabilities, often highly uncertain.

Due to the "trans-scientific" nature of risk analysis, there will always be disputes about methods, end points, and models. Individual and societal values may not be separable from the quantitative analysis, determining what we choose to analyze. Tension over the use of quantitative analysis will be amplified by distributions of gains and losses, as well as prior commitments. Key goals of the risk analyst include extracting the good data from the bad, deciding which model best fits both the data and the underlying process, as well as understanding the limitations of available methods.

In some ways, risk analysis is a mature field, and a number of methods and techniques have become institutionalized. Yet in many profound ways, risk analysis remains immature. To some, the subject amounts to many fascinating case studies in search of a paradigm! The risks of contracting human immunodeficiency virus, of acquiring cancer from pesticides, of nuclear accidents, or of space shuttle disasters are regarded as important but idiosyncratic cases. To the extent that generalized lessons are not learned, science, technology, and environmental policy research has yet to find a common language of expression and analysis.

Despite a number of attempts to rationalize the use of risk analysis in the policy process, its role continues to be controversial. A 1983 National Resarch Council[1] (NRC) project, *Risk Assessment in the Federal Government*: *Managing the Process*, generally referred to as the "Red Book," sought to establish a risk assessment paradigm in the environmental context. It envisioned a sequence of Hazard Identification, followed by parallel Exposure and Dose-Response Evaluations, which are then combined to generate a Risk Characterization. Under this paradigm, once the hazard has been characterized, it can be used to inform risk management.

This approach embodies a technocratic philosophy promoting quantitative risk analysis as *the* solution to arbitrary and "irrational" risk policy decisions. Before he became a Supreme Court Justice, Stephen Breyer wrote in *Breaking the Vicious Circle* that risk assessors should be given an insulated, semi-autonomous decision-making role. John Graham, director of the Harvard Center for Risk Analysis, has campaigned similarly for rigorous training of risk assessors and a central federal department for risk assessment. Legislation that would have mandated quantitative risk assessment for all federal environmental, health, and safety regulations came close to being passed three times in the 1990s:

---

[1]The National Research Council is the research wing of the National Academy of Sciences.

SB110 in 1992, HR9 in 1994, and the Johnston–Robb Bill in 1996. One notable law that did pass eliminated the long-standing Delaney Clause, which had prohibited any known carcinogen as a food additive, regardless of the magnitude of the risk posed by that carcinogen. Under the new law, some levels of carcinogens may be acceptable.

Proponents of a participatory philosophy argue that risk analysis remains too subjective, and its implications too dependent on social context, to permit its removal from the public arena. Since decisions about values and preferences are made not just at the final decision stages but throughout the process, risk analysis necessarily combines both technical expertise *and* value choices. The implications of this interplay range from the inadvertent, as analysts make choices they believe are best without input from interested parties, to the antidemocratic, when the value decisions as well as the number crunching are intentionally restricted to a select group with a particular agenda.

While the "Red Book" approach has come to dominate the way the U.S. Environmental Protection Agency (EPA) approaches risk assessment, many feel that the firewall between "assessment" and "management" is artificial and distortional. Subsequent studies by the NRC (1994 and 1996) begin to address this issue. *Understanding Risk* (NRC 1996) identifies three "outstanding issues": inadequate analytical techniques, fundamental and continuing uncertainty, and a basic misconception of risk characterization. The study concludes that risk analysis must be decision driven and part of a process based upon mutual and recursive analytic-deliberative efforts involving all "interested and affected parties." While clearly more robust and appropriate than an artificial segregation of risk analysis steps, implementation of the *Understanding Risk* approach faces both political and practical obstacles.

Sheila Jasanoff is one of the most vocal proponents of broader representation in risk decision making. Her 1994 critique of Breyer's book (Jasanoff 1994), decrying its artificial separation of fact and value in the risk analysis process, points out that most risk decisions are "far too multidimensional to warrant quantification and much too complex to be simulated through any existing computer program." Jasanoff's view is consistent with the recent National Research Council (1996) review of risk analysis philosophy, which argues for eliminating the misleading firewalls between the assessment and management phases. Regardless of where one fits in these debates, a thorough knowledge of current methods is the vital precondition for effective risk analysis.

"Risk comparison," an approach that has been popular in the past decade can be used to exemplify the two risk philosophies. In a technocratic approach, diverse risks are converted to a common metric —perhaps years of life expectancy lost. Risks are ranked along this dimension, and resources are committed to reducing the greatest risks first. Risk comparisons can also, however, be used as a tool to bring together decision makers to discuss how they perceive risks, evaluate the data available to describe those risks, identify the issues upon which they agree and disagree, and decide when decisions can be made and when more information would be useful.

In 1987 and 1990, respectively, the U.S. EPA and its Science Advisory Board used a technocratic approach to review the ways in which environmental risks were prioritized by the existing regulatory legislation and agencies. These studies found that the existing regulations were inconsistent with both expert and lay opinions of the most important risks. Among the reasons for this inconsistency is that environmental regulation evolved piecemeal in response to individual crises, and over several decades. As a result, regulations use disparate approaches for dealing with different media (air, water, foods, facilities). Some statutes call for absolute levels of safety, some require only "prudent" margins, others base standards on current technology, and some require the regulator to balance risks and benefits explicitly. The reports suggested that the EPA's prioritization should be based more explicitly on risk analysis, but absent legislation specifically allowing intermedia risk comparisons, the EPA's options are constrained by existing laws.

Society compares and ranks risks all the time, although often qualitatively and/or implicitly. In a provocative paper, Wilson (1979) asks risk analysts to make some of these comparisons quantitative. Using a one in a million level of risk (where facing a hazard subjects one to a 0.000001 increase in chance of death from that hazard), Wilson compared some everyday and some less common risks. This sort of simple comparison can be eyebrow raising and may usefully question the wisdom of regulating one risk into oblivion at great cost while far larger risks remain unaddressed; however, such point comparisons are limited and highlight the inextricable nature of value judgment.

Table 1-1 indicates that traveling six minutes by canoe is "equal" along this one dimension to living 150 years within twenty miles of a nuclear power plant. But what does, or can, this comparison mean? There is no indication of the certainty associated with the estimates, the (potentially) offsetting benefits, or ways in which they can be avoided. It

Table 1-1 Risks that Increase Chance of Death by 0.000001 (One in One Million, or 10^{-6})

Smoking 1.4 cigarettes	Cancer, heart disease
Drinking 1/2 liter of wine	Cirrhosis of the liver
Spending 1 hour in a coal mine	Black lung disease
Spending 3 hours in a coal mine	Accident
Living 2 days in New York or Boston	Air pollution
Traveling 6 minutes by canoe	Accident
Traveling 10 miles by bicycle	Accident
Traveling 300 miles by car	Accident
Flying 1000 miles by jet	Accident
Flying 6000 miles by jet	Cancer caused by cosmic radiation
Living 2 months in Denver	Cancer caused by cosmic radiation
Living 2 months in average stone or brick building	Cancer caused by natural radiation
One chest X-ray taken in a good hospital	Cancer caused by radiation
Living 2 months with a cigarette smoker	Cancer, heart disease
Eating 40 tablespoons of peanut butter	Liver cancer caused by aflatoxin B
Drinking Miami drinking water for 1 year	Cancer caused by chloroform
Drinking 30 12 oz. cans of diet soda	Cancer caused by saccharin
Living 5 years at site boundary of a typical nuclear power plant in the open	Cancer caused by radiation
Drinking 1000 24 oz. soft drinks from recently banned plastic bottles	Cancer from acrylonitrile monomer
Living 20 years near PVC plant	Cancer caused by vinyl chloride (1976 standard)
Living 150 years within 20 miles of nuclear power plant	Cancer caused by radiation
Eating 100 charcoal broiled steaks	Cancer from benzopyrene
Risk of accident by living within 5 miles of a nuclear reactor for 50 years	Cancer caused by radiation

Source: Wilson 1979.

is often not even clear that such benefits can be calculated. The real insights, and the real work, come from analyses that address the shape and variability of the risk distributions, the confidence associated with each estimate, and the uncertainty generated by data limitations. Until risks are well characterized, it is difficult even to begin comparing.

The environment is by no means the only arena in which risk analysis is receiving increased attention. As the energy sector is deregulated, risk tools have evolved to deal with variability and uncertainty in supply, demand, pricing, and facility design. Similarly, the rash of major catastrophes in the past several years, including seasonal wildfires and major earthquakes in the west, flooding in the midwest, and Hurricane Andrew and beach erosion in the east, has forced the government to incur large expenses, prompted concern about the viability of private insurance underwriting, and promoted more careful risk exposure assessment. As a final example, increased reliance on information technologies has generated concern among public and private decision makers over the security and stability of computer networks. The set of risk analytical tools presented in this volume may be applied to any of these issues.

Cost-benefit analysis (CBA), a version of decision analysis, increasingly accompanies risk analysis on the public policy agenda. Some critics see CBA as nothing more than risk analysis made more complex by adding value judgments such as those putting dollar values on illness, loss of lives, or degradation of ecological resources. (See Costanza et al. 1997 for an example of a truly "grand scale" economic analysis—the value of global natural resources to society—that generates an estimate at the cost of massive uncertainty.) In some cases, simply listing all relevant impacts (positive and negative) without absolute valuations will provide insight into a decision. In other cases, optimizing costs and benefits requires the analyst to quantify all of the tradeoffs in a common metric, usually monetary values. If these choices are, in fact, incommensurate, forcing dollar values on them may be at best arbitrary and at worst self-serving.

Others see CBA as a tool that, while fraught with uncertainty, gives a common rule by which to make necessary comparisons. They argue that society makes these comparisons already, and that CBA will do so in a more consistent, rational manner. The 1994 bill HR9, would have made CBA a legislative requirement, requiring "a final cost-benefit analysis" for every "major rule." Figure 1-1a (after Morgan 1981) represents an idealized economist's approach to solving this problem. However, for many risk issues, the values of the different elements range from extremely difficult to impossible to quantify. How, for example, can we measure the value sports enthusiasts place on the opportunity to play outdoors, and compare this to the costs they impose on society through skin cancer treatment? Figure 1-1b may better represent what we know about many risks.

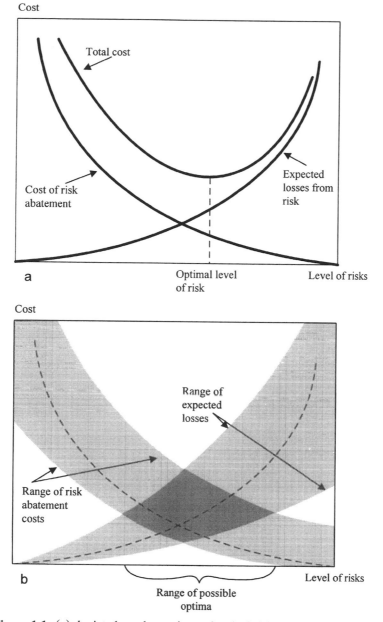

Figure 1-1. (a) depicts how the optimum level of risk and abatement can be calculated, given precise information on costs, benefits, and preferences. (b) suggests that, even if preferences are clearly defined, uncertainties in risk and abatement costs can lead to highly uncertain ranges of possible optima.

In 1985, Professor John Harte of the University of California at Berkeley's Energy and Resources Group created a course on environmental problem solving. Harte's approach was to equip students with a few general tools that allow them to address problems characterized by limited information and apparent complexity. He teaches his students "an approach to problem solving [that] involves the stripping away of unnecessary detail, so that only the essentials remain" (Harte 1985).

Harte presents a three-step approach, a philosophy that he spells out in the preface to *Consider a Spherical Cow* (Harte 1985, pp. xi–xiii). First, he takes a broad overview of a problem (what he calls hand-waving), in order to establish a qualitative understanding of the mechanism of the process being examined. Looking at the "big picture" can often provide an idea of the direction and magnitude of a process, even if the details are obscure. In addition, it can quickly become evident where important information is missing, and which assumptions are most problematic. At this stage, simple "reality checks" suggest whether the solver is on the right track.

Second, he represents the qualitative processes mathematically and uses available data (making assumptions where necessary) to arrive at a "detailed quantitative solution." Third, he evaluates the resilience of his answers if the assumptions he has made are changed or omitted. This step, also called sensitivity analysis, can be applied to both data and assumptions, to suggest where further research will improve understanding and whether uncertainty about the assumptions is likely to overwhelm the results.

In this book, we adapt and extend Harte's environmental problem-solving approach to risk analysis. Harte's philosophy is wonderfully appropriate to risk assessment, where uncertainty is often profound and assumptions must inevitably be made. To familiarize the reader with hand-waving techniques, the following three problems consider important risk problems without using any numbers.

Problem 1-1. Getting Started

Consider figures 1-2 through 1-5. Which of these graphs do you think best represents

  a. the number of accidents a driver has, as a function of total cumulative miles driven?

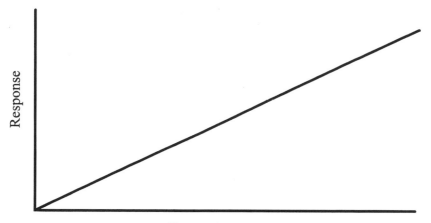

Dose or exposure

Figure 1-2. A direct relationship between a dose (cause) and its response (effect). It need not be one to one, but it must be true that a unit increase in dose causes a constant increase in response. An example is the purchase of raffle tickets: the chances of winning are directly proportional to the fraction of the total tickets that you hold.

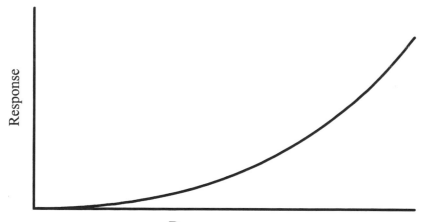

Dose or exposure

Figure 1-3. A convex relationship, where increases in dose have a relatively larger impact than initial dose. The risk of highway accident per mile traveled as a function of travel speed is a convex relationship. Convex functions are those for which the second derivative is positive, meaning that the slope of the line increases throughout the convex region.

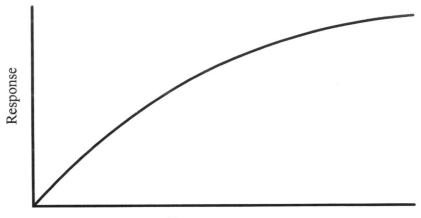

Figure 1-4. A concave relationship where additional dose has a smaller relative response than does the initial dose. This curve describes, for example, the relationship between pedaling effort and bicycling speed: wind resistance increases at a faster rate than the increase in speed. Concave functions are those for which the second derivative is negative, meaning that the slope of the line decreases throughout the concave region.

    b. the number of space shuttle accidents as a function of total number of missions flown?

    c. the number of leaks in a sewer line as a function of the number of years it has been in service without maintenance or replacement?

    d. the number of carcinomas a surfer is likely to get as a function of total lifetime hours in the sun?

Solution 1-1

Differing interpretations are possible for several of the cases, and without further analysis we are, at this point, making educated guesses. However, there can often be considerable value to these educated or hand-waving guesses. It is important not to wave your hands frantically like a lost hiker hoping to be spotted by a passing plane, but rather in the controlled and directed fashion of a symphony conductor. Thinking in very general terms can often point out where we have good enough information to make a reasonable decision, and where our guesses are so broad or unrealistic that we must push the questions further.

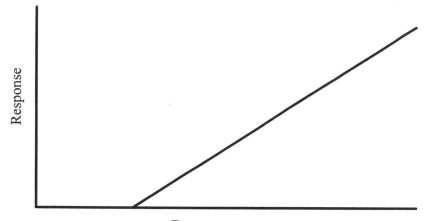

Figure 1-5. A threshold relationship in which initial doses have no effect, but eventually a dose (the threshold) is reached beyond which a response is elicited. An example is a redundant system, where the first several failures are protected by backups, but eventually the backups are exhausted and effects begin to appear. Note that threshold effects can include linear, convex, or concave patterns.

Solution 1-1a

Since drivers tend to gain experience over time, figure 1-3 (concave) is probably a good representation. The driver's total number of accidents will grow, but at a decreasing rate. In this context, the concave response is often referred to as a "learning curve," as the failures grow less frequent with experience, and there are diminishing marginal benefits from experience. In some physical contexts this curve describes saturation, where subsequent doses do not elicit as much response.

Solution 1-1b

Assuming that routine maintenance is done, and the launch- and space-worthiness maintained, figure 1-2 (linear) might be a good choice (see point one below). There is a roughly constant probability of an accident for each launch, which leads to a linear cumulative hazard.

Solution 1-1c

One would expect a new sewer line to have a fairly high integrity, which would suggest a very small number of leaks early on. However, as it continues in service without maintenance, ground shifting, corrosion, and other effects cause increasing numbers of leaks, and probably at an increasing rate. Consequently, figure 1-4 (convex), an exponentially increasing number of leaks over time, may be a good model, or possibly figure 1-5 (threshold), if it takes a while for the first leak to start, and then the effects of the corrosion from the initial leaks exacerbate the effects of other deleterious forces.

Solution 1-1d

There are arguments for any of the four models here; the same is true for many causes of cancer and other types of chemical toxicity. Figure 1-2 makes sense if each unit of energy is equally likely to create a cancer cell. Figure 1-3 would apply if there were some saturation effect where most of the damage is due to the initial exposure to radiation. Figure 1-4 implies that additional radiation is likely to exacerbate the effect of earlier exposure, meaning an increasing rate of carcinomas over time. Finally, figure 1-5 (threshold) represents the case where the body is able to repair the damage due to a limited amount of radiation (or, for example, to metabolize a toxin up to some amount), but beyond that threshold level, it cannot, and carcinogenesis begins. The question of whether a process has a threshold is fundamental to quantitative and qualitative risk analysis.

Problem 1-1 highlights several issues. First, there is often insufficient information to come up with an indisputably "right" answer. We make a number of assumptions about what is causing the phenomena we are interested in and use these assumptions to theorize about the expected outcome. Depending on what aspects we think are most important and relevant, we may individually arrive at quite different sets of assumptions. For example, you may have assumed that there is a learning curve associated with space shuttle launches, in which case the curve would be concave. On the other hand, if you thought that the individual shuttles were likely to be subject to wear and tear, you might predict an exponential increase in the number of accidents.

Second, based on our theories about how things work, we "build models." Throughout this book, you should keep in mind that the concept of modeling is simple: use what we know to describe what we observe. The models in this exercise were built without using equations or numbers, but we have an idea of what is on each axis and how the axes relate. Plugging in the numbers (provided we can get them) and calculating can often be a trivial exercise; the important point is to understand what is going on. The benefit of eventually plugging in numbers is that we can use them to predict future outcomes. While this ability to predict is the goal of risk assessment, differences in assumptions and theories can lead to highly divergent numbers, that is to say, uncertainty.

Third, a single model can be used to describe very different phenomena. The essential modeling relationships of, for example, carcinogenesis and automobile accidents may be analogous, even when the physical processes are entirely different. This is extremely important, because it allows us to develop general methods for thinking about a wide range of problems. The next step is to refine the models and make them better fit the specific case under scrutiny. This in turn requires more information.

Problem 1-2. Data Needs

What evidence would you want to confirm your (or our) answers to problem 1-1?

Solution 1-2

Problem 1-1 is about constructing theoretical models; problem 1-2 is about verifying and calibrating the models empirically by comparing them to data. Two things to keep in mind are how well the data fit the model and how "good" the data are ... and recall that bad data may erroneously "confirm" a bad theory!

Solution 1-2a

The insurance industry has an abundance of data on this subject. In general, younger drivers tend to get into more accidents than do older drivers, but the decreasing trend tends to plateau at some age over thirty. Note that there may be two mechanisms operating here. One is experience—the number of years that an individual has been driving, and the skills he or she has gained through that practice. The other is maturity—older drivers may be less risk-taking relative to younger drivers. Note also that insurance companies usually use individual data as well, such as number and magnitude of prior accidents.

Solution 1-2b

Richard Feynman, in a 1988 article documenting his review of the space shuttle Challenger 1985 explosion, found that some of the engineers estimated about a one in two hundred chance of such a failure, based on their understanding of the materials and very complex equipment involved. Meanwhile, people at higher levels in the administration assumed much smaller probabilities, on the order of one in ten thousand. The accident occurred on the seventy-eighth flight, and while limited inference is possible given only one occurrence, the engineers' model appears better supported than that of the administration. Why the difference? It is likely—as the subsequent investigation showed— that the politics and finances of the shuttle program exerted a strong pressure to remain "on schedule." Thus, while more information would improve decision making about this particular risk, we are not likely to get it in time to make good decisions. (In fact, while additional safe flights extend the data set, we hope we do not get additional "failure" data!) Consequently, the choice of the right model will be based on the extent to which we believe the assumptions behind each. The moral of the story is that not all assumptions can be tested. A large number of failures makes modeling easier, while few (or no) failures makes predictions extremely uncertain.

Solution 1-2c

Data on this could come from a variety of sources. Ideally, one would want to inspect the pipe in question regularly and check for leaks. Alternatively, a comparison to similar pipes in similar use might provide relevant information. Laboratory tests on the pipe may give information about the susceptibility to failure over time, and geologic history might suggest the types of stresses the pipe is subject to. Manufacturers and sewer companies may have historical and laboratory data on material specifications.

Solution 1-2d

The types of data needed to support one answer or another come in two broad classes: toxicological and epidemiological. We will consider both of these in much greater detail later in the book. In general, a toxicological test would involve exposing groups of individuals to varying levels of sunlight (or ultraviolet light), while keeping everything else in their lives the same, to see whether different skin cancer levels result. Since it is difficult (and ethically unacceptable) to do this sort of test on people, it is more often done on small groups of animals or cultured human cells. Additional assumptions must therefore be made, for example, about the relationship between animal and human carcinogen susceptibility.

An epidemiological test would try to find individuals who have been exposed to different levels of sunlight in the past, and compare rates of skin cancer among those groups. The problems here are likely to be with data quality, such as accuracy in determining past exposure, and in insuring that some third factor that was not measured is not the cause.

Problem 1-3. Using Data

Assume that you found that for problem 1-1d, figure 1-3 was the most likely model, and that 50% of people who surfed regularly could be expected to get at least one carcinoma. How would this affect your attitude about surfing? How would you tell people about your findings? Would you suggest that surfing be regulated?

Solution 1-3

First, note that this issue will come up again much later in the book, when evaluating the relationships between numbers, such as one in a million, 1%, 10%, 50%, and more abstract concepts, such as "rare," "very rare," and "common," and what these distinctions and relationships mean when it comes to making policy.

Clearly, there is no obvious answer to this question. However, where the first two problems asked us to formalize our thinking about risk, this one asks what to do once we have information. We need to think about how people perceive risks—does the average surfer think she has a 50/50 chance of getting cancer? We need to think about how to advise surfers about their risks. We also need to think about how and whether to try to compare this cancer risk to other risks and benefits, how and whether surfing fits into our social and regulatory system, and whether and what additional information would improve our decisions. Finally, we need to think about which surfers are at risk. Does each of them have a 50/50 chance to contract cancer, or are some of the surfers at higher risk for genetic or other reasons? If people do vary, can we figure out which are the high-risk surfers? And if we can determine the high-risk individuals, can and should we treat them differently? Finally, what actions can surfers take?—sunscreen, wet suits, T-shirts, and so on.

Problem 1-A. Additional Cases

Answer problems 1-1 and 1-2 for the following cases:

a. The expected number of "heads" from tosses of a fair coin as a function of total number of tosses
b. The probability of a ski jumper crashing as a function of the height of the mogul from which she jumps
c. The number of times a cheap handgun will misfire as a function of the number of times it is fired

Problem 1-B. Additional Curves

a. Consider the observation that at some stage in their lives, the competence of many drivers begins to deteriorate. Draw a curve that represents a driver's lifetime driving experience, beginning with a steep but gradually leveling learning curve, followed by a long period of no change, followed by increasing risk late in life.

b. For some processes and items—washing machines, for example—there is some chance that the item will fail upon initial use. However, if the item operates successfully the first time, the additional chance of failure grows very slowly over time. Graph this phenomenon.

c. Suggest some additional curves, along with cases that they might represent.

Problem 1-C. Does the Dose Make the Poison?

a. Assume that newborn body weight is a reasonable measure of health, with higher weight meaning better health, and that vitamin D is essential in moderate quantities but injurious in excess; implying an optimum dose corresponding to a maximum average birth weight. Propose a curve that would represent newborn body weight as a function of the mother's intake of vitamin D.

b. Repeat (a) using reduction in body weight instead of body weight. How do the curves differ? In what respects are they the same?

c. Rephrase (a) such that it asks the same question but is represented by a (seemingly) different curve. Discuss the effects the different representations might have on perceptions of the effects.

Problem 1-D. One in a Million Risks

Refer back to table 1-1.

a. Why are there two different entries for risk from coal mines? How would this table seem different if the two were combined into a single 10^{-6} risk of spending three quarters of an hour in a coal mine?

b. According to this table, if you lived in Miami and drank a can of diet soda each week, your risk from that would be greater than that from drinking the tap water. Does this mean that you are unreasonable if you object to carcinogens in your drinking water?

That you are taking on too much risk from diet soda? Neither?
Discuss.

Problem 1-E. Surfing and Smoking

The debate about managing cancer and other risks associated with
cigarettes is both similar to and dissimilar from that about skin cancer
and other risks from surfing. Compare the question of public health and
surfing to that of public health and smoking. In what ways do the issues
differ, and in what ways are they the same?

Problem 1-F. Risks of Nuclear Power

The operation of large (1-GW_e, or 10^a watts, scale) nuclear power
reactors began in about 1970 and there are approximately 350 nuclear
power reactors operating worldwide today.

a. Roughly how many nuclear power reactor-years of operation have
   accumulated during this period?

b. During this period, there has been one accident that resulted in a
   major release of radioactivity (Chernobyl 1986) and one accident
   in which all but a small amount of release was prevented by the
   reactor containment building (Three-Mile Island 1979). On this
   basis, roughly what is the probability of a major release per
   reactor-year?

c. In 1975, the WASH-1400 Reactor Safety Study Report (see box 1-2
   and figures 1-6 and 1-7) estimated that the chances of a Cher-
   nobyl-type accident were around one in one million each year. Is
   this consistent with your estimate in (b)? What might account for
   the differences?

d. The Chernobyl accident may ultimately cause on the order of
   10,000 extra cancer deaths (von Hippel and Cochrane 1991). How
   many would this come to per reactor-year?

Box 1-2. The WASH 1400 report

In 1975, the United States Nuclear Regulatory Commission completed a now famous (or infamous) report entitled "Reactor Safety Study," which tried to quantify the probability of various types of reactor accidents that might occur, and compare those risks to other risks that people already face. The report continues to be discussed because it generated both great applause and criticism—sometimes both from the same individuals! The report's proponents argue that it was the most thorough and quantitatively rigorous risk analysis ever done in any context, and provided the best possible numbers to inform policy makers. Its detractors counter that it completely ignored potentially disastrous interactions, and used comparisons (such as tornadoes versus core meltdowns) that ignored inherent differences.

In addition, while including immediate radiation-induced deaths caused by various possible failure scenarios, figures 1-6 and 1-7 failed to depict associated long term cancer deaths. As in the Chernobyl case, the latter could be orders of magnitude greater than the number of short-term deaths (von Hippel and Cochran 1991). Consequently, the curves on figures 1-6 and 1-7 are highly misleading. For example, a curve taking into account long-term cancer deaths would predict about 10,000 deaths at a frequency of 10^{-3} per year for 100 reactors, rather than the figure 1-6 and 1-7 predictions, which approaches zero fatalities at that frequency! (Hohenemser et al. 1992).

---

   e. Each year in the United States there are roughly fifty fatalities in coal mines, a few hundred in coal transport, and a few thousand due to respiratory diseases caused by the emission of SO_2 by the equivalent of three hundred 1-GW_e U.S. coal-fired power plants. On this (quite uncertain) basis, approximately how many coal-fired power plant fatalities are there per equivalent reactor-year?

   f. Society does not respond in the same way to all risks, for example, the risks from nuclear and coal power plants—or to the fatalities from auto accidents, smoking, and skiing. Discuss why this appears to be the case.

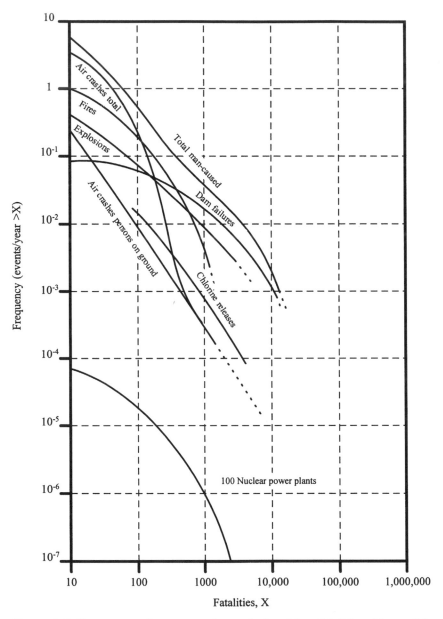

Figure 1-6. Frequency of man-caused events involving fatalities. Figure 6-1 from WASH-1400 report (U.S. Nuclear Regulatory Commission 1975).

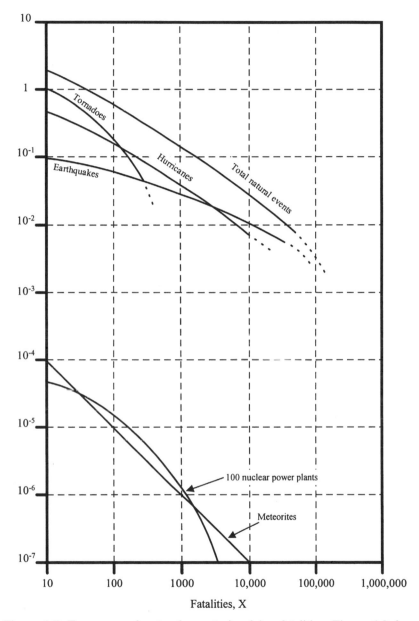

Figure 1-7. Frequency of natural events involving fatalities. Figure 6-2 from WASH-1400 report (from U.S. Nuclear Regulatory Commission 1975).

References

Breyer, S. G. (1993). *Breaking the Vicious Circle: Toward Effective Risk Regulation*. Cambridge, MA: Harvard University Press.

Costanza R., d'Arge, R., de Groot, R., Farber, S., Grasso, M., Hannon, B., Limburg, K., Naeem, S., O'Neill, R., Paruelo, J., Raskin, R, Sutton, P., and van den Belt, M. (1997), "The value of the world's ecosystem services and natural capital." *Nature* 387:253–60.

Covello, V. T. and Mumpower, J. (1985). "Risk analysis and risk management: An historical perspective," *Risk Analysis* 5(2):103–20.

Feynmann, R. P. (1988), "An outsider's inside view of the Challenger inquiry," *Physics Today* (February) 26–37.

Harte, J. (1985). *Consider a Spherical Cow: a Course in Environmental Problem Solving*. Mill Valley, CA: University Science Books.

Hohenemser, C., Goble, R., and Slovic, P. (1992). "Nuclear power." In Hollander, J. M. (Ed.) *The Energy-Environment Connection*. Covello, CA: Island Press.

Jasanoff, S. (1993). "Bridging the two cultures of risk analysis." *Risk Analysis* 13(2):123–29.

———. (1994). "The dilemmas of risk regulation." *Issues in Science and Technology Policy* (Spring):79–81.

Morgan, M. G. (1981). "Choosing and managing technology-induced risk," *IEEE Spectrum* 18(12):53–60.

National Research Council. (1983). *Risk Assessment in the Federal Government: Managing the Process*. Washington, DC: National Academy Press.

———. (1994). *Science and Judgment in Risk Assessment*. Washington, DC: National Academy Press.

———. (1996). *Understanding Risk*. Washington, DC: National Academy Press.

Otway, H. (1992). "Public wisdom, expert fallibility." In Krimsky, S. and Golding, D. (Eds.) *Social Theories of Risk*. NY: Praeger.

U.S. EPA (1987). *Unfinished Business: A Comparative Assessment of Environmental Problems*. Washington, DC: U.S. EPA Office of Policy Analysis.

———. (1990). *Reducing Risk: Setting Priorities and Strategies for Environmental Protection*. Washington, DC: U.S. EPA Science Advisory Board.

U.S. Nuclear Regulatory Commission. (1975). *Reactor Safety Study: An Assessment of Accident Risks in U.S. Commercial Nuclear Power Plants*. Executive Summary, WASH-1400, NUREG-75/014.

von Hippel, F. N. and Cochran, T. (1991). "Chernobyl: estimating the long term health effects." In *Citizen Scientist* New York, NY: Touchstone Books.

Weinberg, Alvin. (1972). "Science and trans-science." *Minerva* 10(2):209–21.

Wilson, R. (1979). "Analyzing the daily risks of life." *Technology Review* 81(4):41–46.

2

Basic Models and Risk Problems

Introduction

Consider the following three situations:

(1) Over the past decade, the United States Department of Energy (DOE) has been trying to site a high-level radioactive waste (HLRW) depository at Yucca Mountain in Nevada. Currently, as has been the case for decades, nonmilitary HLRW is stored where it is generated, primarily at nuclear power plants. If built, Yucca Mountain will be a giant tunnel, twenty-five feet in diameter and thousands of feet deep. It must be designed to contain HLRW for ten thousand years, the amount of time it takes to become "safe." Suppose you are retained as a consultant to advise the DOE, and asked to look into the following. How might you proceed?

- What is the risk of human exposure from the site over the next ten thousand years?
- What can be done to prevent such exposure?
- If estimates provided by the best available earth scientists vary by several orders of magnitude, how can we estimate the probability of future volcanic activity in the area?

(2) A number of toxicological and epidemiological studies suggest that benzene is a carcinogen. Table 2-1 shows the results from a laboratory animal test, and table 2-2 those from an epidemiological study. You have been asked to determine the likelihood that someone exposed to benzene in the workplace over thirty years will get cancer as a result. How would you address these issues:

- What is the relationship between animal and human metabolism, and what does it tell us about carcinogenesis?
- What assumptions must we make about the mechanisms of cancer when we extrapolate from high to low doses?

Table 2-1 Benzene Exposure and Tumors from Laboratory Test

Number of test rats	Number of rats with tumors	Rat test dose (mg/kg/d) (Oral gavage)
50	0	0
50	4	25
50	20	50
50	37	100

Source: Crump and Allen 1984.

- How do we determine whether cancers in a group of workers can be attributed in part to benzene?

(3) Consider the information in table 2-3, which summarizes accident data for major U.S. airlines, as of 1993. You are asked to rank the airlines according to how safe you believe them to be, and to estimate risks from future flights on each airline. How would you proceed?

Certainly, we don't expect you to be able to answer these questions right now. By the end of the book, however, you should be able to establish and evaluate the major parameters for each issue, identify how the problem can be addressed, evaluate what information you have and what you still need, and characterize the uncertainty associated with possible solutions. Consistent with the "walk before you can run (but then run)" philosophy of this book, we encourage you to spend a few minutes thinking about these three problems now, and save your work. Then return to them as you go through the book, so that you can track the tools you have learned. Finally, write up answers after you have worked through the book and see how those answers compare to your original efforts.

Basic Modeling

In the last chapter, we introduced a number of risk assessments that we will now revisit in a more systematic fashion. Figures 1-2 to 1-5 in the last chapter demonstrate how very different phenomena can be represented by the same sort of figure. This chapter will look at two major elements that are consistently useful: stock-flow models, which provide

Table 2-2 Benzene Exposure and Observed Incidence of Cancers by Type

Cumulative exposure (ppm-years)			Total lymphatic and hematopoietic		Leukemia		Total excluding leukemia	
Range	Mean	Person-years	Observed deaths	Expected deaths per person-year	Observed deaths	Expected deaths per person-year	Observed deaths	Expected deaths per person-year
0–45	11	30482	6	2.02×10^{-4}	3	7.91×10^{-5}	3	1.23×10^{-4}
45–400	151	16320	6	2.35×10^{-4}	4	9.19×10^{-5}	2	1.43×10^{-4}
400–1000	602	4667	3	3.39×10^{-4}	2	1.34×10^{-4}	1	2.03×10^{-4}
>1000	1341	915	6	4.81×10^{-4}	5	1.97×10^{-4}	1	2.95×10^{-4}
Total	132	52584	21	2.30×10^{-4}	14	9.03×10^{-5}	7	1.40×10^{-4}

Source: Crump 1996.

Table 2-3 Accident Data for Major U.S. Airlines

U.S. airlines	Death risk per flight
U.S. Air (US)	1 in 2.5 million
Northwest (NW)	1 in 4 million
TWA (TW)	1 in 5 million
United (UA)	1 in 8 million
American (AA)	1 in 10 million
Delta (DL)	1 in 10 million
Continental (CO)	1 in 15 million
Southwest (WN)	0 (perfect record)
Industry average	1 in 7 million

Source: Based on data from a study of total fatalities over a twenty-year study period (*Newsweek* 1994).

a universal tool from which to consider most exposure scenarios, and cause-effect models, which can be used to predict the effects given the exposure.

The primary goal of this chapter is to accustom the reader to approaching risk problems by looking first for the key elements and most appropriate tools. Only after these have been identified is more sophisticated and detailed analysis appropriate. In addition, these problems are intended to demonstrate how much information and understanding can be gleaned from what are often very simple models. Conversely, poorly conceived or misapplied models can obscure even the simplest system.

This book refers to models extensively; it is a ubiquitous piece of the risk jargon. While the concept of mathematical modeling may seem intimidating, in reality it need not be. A model is a simplified description of how things in the world work. By building models, we gain insights into what outcomes can be expected under specific conditions, and can simulate what might happen if we change assumptions and parameters.

It is important to have a healthy disrespect for models. Any interesting or insightful results they produce result from the initial formulation...whether it is correct or not! A model is a *vehicle* to help guide insight. Too much time spent modeling, or taking models themselves as a form of truth, is not only counterproductive, it can be dangerous.

Problems 2-1, 2-2, 2-A, and 2-B are based on stock-flow models.

Problem 2-1. Volatile Organic Emissions from Household Materials: Wallpaper Glue

Many consumer products release chemicals during the first few days to years of use. For example, wallpaper glue can "leak" 1-1-1 trichloroethane (TCA), a ubiquitous manmade chemical. Assume that you have papered the walls of a 10 by 15 foot shed with 8 foot high walls. The wallpaper glue releases TCA at the rate of 13 μg of TCA per square foot of plywood surface area per minute (μg/ft^2/minute). What is the equilibrium concentration of TCA in the room? (Assume that the air inside is replaced with air from the outside twice each hour.) Assume for the purposes of this problem that the TCA flow rate is not dependent on concentration in the building (as a supplemental problem, include this effect).

Note: for further data on some volatile organic emissions from household materials, see Wallace et al. 1987.

Solution 2-1

This problem presents a quantity, or "stock," of TCA stored in a "reservoir" of wallpaper glue, and the TCA "flows" at a known rate. Stock-flow problems have a number of uses in risk analysis. They can give us information on exposures to hazards (for example, inhaled TCA is known to be carcinogenic), and can indicate which sources are important, and whether sources or removal pathways are more important as mitigation opportunities, that is, where we can take risk-reducing measures.

The first step in solving a problem like this is to structure it in a way that makes sense. Often, a useful first step is to diagram the process (fig. 2-1). Equations can then be used to describe the diagram, and once the equations are produced, numbers can be input and solved. Keep in mind that describing the problem efficiently and accurately is both

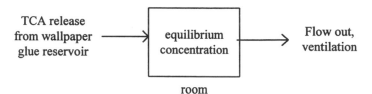

Figure 2-1. Line-box diagram of TCA stocks and flows for problem 2-1.

important and challenging, while finding the final numerical solution is just a matter of "cranking the numbers."

The key approximation to make when solving this problem is that the concentration of TCA has reached a "steady state" inside the house. That is, the TCA concentration is independent of time. This assumption breaks down for data that are collected over sufficiently small time intervals because a number of factors that affect the ventilation of the house and the rate of formaldehyde entry fluctuate about their average values on certain time scales. (The number of opened windows affects the rate of air exchange, for instance, as do outdoor and indoor temperatures.) Additional variation is found at the longer, seasonal time scale; a room may be ventilated more in the summer than in the winter.

Nonetheless, it makes sense that this approximation holds over a long time period, because otherwise the TCA concentration would either dwindle to zero or increase without bound. Since neither of these makes sense, let us ignore for the time being the fluctuations in the TCA entry rate and the ventilation rate, and use only their average values.

The flow rate of air is given in terms of air changes per hour (2 ACH), which means that outdoor air enters the room, and therefore indoor air is removed, at the rate of two room volumes every hour. F_{in}, the flow rate of TCA into the shed, can be calculated using the release rate of TCA and the surface area of the walls:

$$F_{in} = (13 \ \mu g/ft^2/minute) \cdot (\text{area of walls})$$

$$= (13 \ \mu g/ft^2/minute) \cdot (2 \cdot 8 \cdot 10 + 2 \cdot 8 \cdot 15)ft^2$$

$$= 5200 \ \mu g/minute.$$

In equilibrium, the flow in will equal the flow out, or

$$F_{in} = F_{out} = 5200 \ \mu g/minute.$$

Assuming that ventilation removes most of the TCA from the shed (breakdown of TCA is a minor removal path indoors), the concentration of TCA in the room is equal to that of the air leaving the room. Using this and the air exchange rate given above, the concentration can be calculated:

$$F_{out} = V_{room} \cdot (\text{air exchange}) \cdot C_{room},$$

so that

$$C_{room} = F_{out}/[V_{room} \cdot (\text{air exchange})]$$
$$= (5200 \ \mu g/\text{minute}) \cdot (60 \ \text{minutes/hour})/$$
$$[8 \cdot 10 \cdot 15 \ ft^3 \cdot 2/\text{hour})]$$
$$= 130 \ \mu g/ft^3.$$

Clearly, this is only a starting point. Why does it matter that there is 130 μg of TCA per cubic foot in the shed? Would breathing this much TCA be fatal in one minute, harmless if inhaled twenty four hours per day, or something in between? Calculating the concentration is evidently only the first step in the process of risk analysis. This calculation required only a balancing of concentrations (i.e., stocks and flows). What if the source of the ventilation is not constant? We will use the next problem to develop some techniques for that situation. Other approaches to these questions are found in later chapters.

Problem 2-2. Indoor Radon Exposure

Radon is a radioactive element found in trace amounts in the atmosphere as the decay product of radioactive materials in the earth. It is generally found at higher concentrations indoors, as air in buildings circulates more slowly than does outdoor air, allowing radon to accumulate. Indoor radon concentration reaches a steady state as increase in radon coming from the ground (and other sources) is balanced by the loss of radon through decay and diffusion to the outside. It is possible to reduce the radon level in a home, and the U.S. Environmental Protection Agency (EPA) uses 4.0 pCi/l as the concentration at which such remedial action should be taken (U.S. EPA 1992).

Table 2-4 Radon in Homes

Typical ventilation rate	0.5 ACH (Air changes per hour)
Annual average indoor radon concentration	1.5 pCi/l
Radon half-life	3.8 days
Rate of radon entry	Unknown (pCi/hour)

Radon doses are measured in picocuries per liter of air, where one curie is the amount of radiation emitted by one gram of radium. Radon is believed to be a carcinogen, or cancer-causing agent, based on a number of epidemiological studies of populations exposed to unusually high levels of radon. This problem uses real data on radon exposure in the United States to begin to illustrate basic statistics. Later in the book, the methods used to derive these values will be addressed, along with how believable they are. This problem assumes that they are valid.

 a. Given the information in table 2-4, what is the average rate at which radon enters a "typical" (2000 square foot) single-story house? Assume that the concentration of radon in outdoor air is negligible.

 b. Calculate the flow rate of radon into a typical house that has an average annual radon concentration of 16 pCi/l. How much additional airflow would be necessary to reduce the concentration in this house to 1.5 pCi/l?

 c. Estimate what would happen to the winter heating bills of a house in which the ventilation rate is increased tenfold, assuming that air exchange accounts for 20% of household heat loss (with the other 80% resulting from conductive cooling through walls and roof). Based on this estimate, is increasing ventilation this much a cheap way to reduce the amount of radon in a house?

 d. Suggest some radon mitigation strategies that might be less expensive.

Solution 2-2

This problem is similar to problem 2-1, except you must now calculate the flow instead of the stock, and add an additional exit route—radioactive decay. Radioactive decay describes the breakdown of one atomic species into another by emitting α-particles (two protons plus two neutrons, i.e., the nucleus of a helium atom) and β-particles (electrons). The decay of radon (or any radioactive isotope) is described by a half-life, which is the amount of time required, on average, for half of the atoms in a sample to undergo radioactive decay. Other exit routes include metabolism (especially of organic compounds) in the body and chemical degradation (for example, formaldehyde exposed to sunlight breaks down).

Before digging into the problem, it is useful to think about why radon in homes is worthy of our attention. Radon (Rn-222) and related compounds are radioactive isotopes that result from the decay of naturally occurring radium 226, which is universally present at low concentrations in soil and rock. Figure 2-2 shows the decay process. In general, when houses are built in geologic formations with relatively high concentrations of radioactive elements, radon can enter the house through the floor or (usually to a lesser extent) via the water supply. While a number of isotopes are involved, they are found in approximately consistent proportions because of the consistency of the radioactive decay process. Consequently, for many purposes the measured level of radon is taken as a proxy for the level of radioactivity due to radon and its decay products.

Radon was identified as a potent carcinogen when uranium miners began exhibiting abnormally high cancer rates. It was found that these workers were exposed to radon concentrations ranging up to thousands of pCi/l. More recently, it has been found that some homes in the United States have quite high naturally occurring levels of radon indoors, leading to the concern that this may be a highly significant cancer hazard. As figure 2-3 shows, there is significant variability among homes across the US. There can even be dramatic differences between neighboring homes.

Living one's whole life in a home containing 4 pCi/l of radon (a little over twice the national average concentration of a house in the United States) is thought to be associated with an increased cancer risk of 10^{-6}. This assessment is based largely on linear extrapolation from individuals exposed to high levels. Furthermore, it is problematic in that it averages across the population, ignoring the expectation that added risk would be higher among smokers, for example.

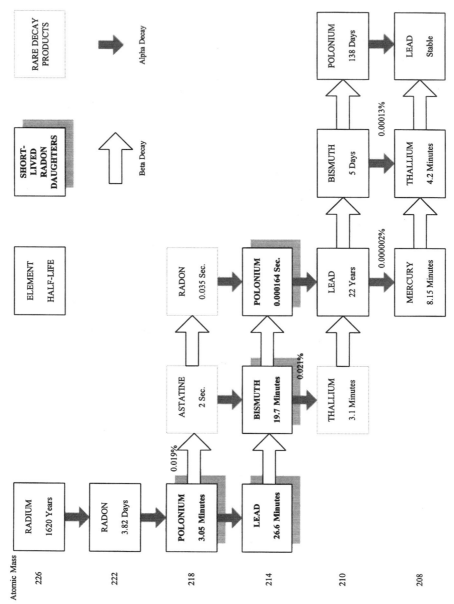

Figure 2-2. Radium decay chain.

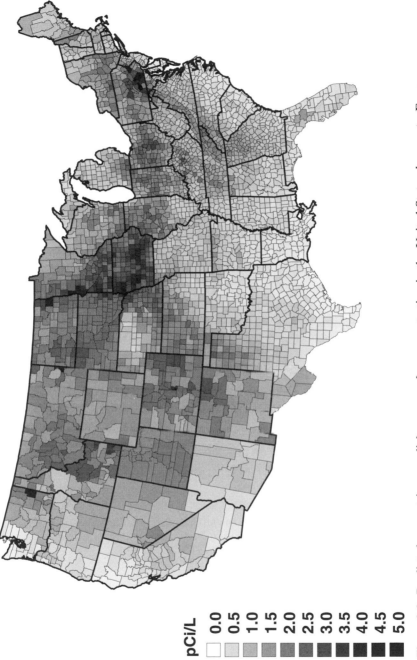

Figure 2-3. Predicted geometric mean living-area radon concentration in the United States, by county. From Price et al. 1998.

Solution 2-2a

There is one flow of radon into the house—this is what we're solving for. There are two flows out—air exchange and radioactive decay. Consequently, we can represent this problem with figure 2-4 and equation 2-1:

Flow in = (flow out, air exchange) + (flow out, radon decay).   2-1

Since the flow out via the air is dependent upon the equilibrium concentration and independent of radon decay, it can be calculated as in problem 2-1. The concentration of radon in the house is known, so the next step is to calculate the flow rate of radon from the house, which requires an estimate of the size of the stock, or air capacity of the whole house. The surface area of a typical house is about 2000 square feet, and standard indoor ceilings are 8 feet high. Therefore the volume of the house is

$$2000 \text{ ft}^2 \times 8 \text{ ft} = 16{,}000 \text{ ft}^3.$$

To convert the units from air changes per hour (ACH) to cubic feet per hour, multiply the ventilation rate by the volume of air in a typical house, or

$$(1 \text{ ACH}) \times (16{,}000 \text{ ft}^3 \text{ air/air change}) = 16{,}000 \text{ ft}^3/\text{hour}.$$

Next, calculate the number of Rn atoms in the house at any given time. A curie is a measure of decays per unit time, and

$$1 \text{ pCi} = 0.037 \text{ decays per second}.$$

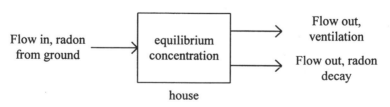

Figure 2-4. Radon stocks and flows for problem 2-2.

However, since decays/per second and number of atoms are both constant, the rate is an appropriate "stock" to calculate.

The amount of radioactivity in the building is

$$16{,}000 \text{ ft}^3 \times \left(\frac{\text{m}^3}{35.31 \text{ ft}^3} \right) \left(\frac{1000 \text{ liters}}{\text{m}^3} \right) \times \left(\frac{1.5 \text{ pCi}}{\text{liter}} \right)$$

$$= 6.8 \times 10^5 \text{ pCi}.$$

Therefore flow out via decays is

$$\left(\frac{6.8 \times 10^5 \text{ pCi}/2}{3.8 \text{ days}} \right) \left(\frac{\text{day}}{24 \text{ hours}} \right) = 3.7 \times 10^3 \text{ pCi/hour},$$

and flow out via air exchange is

$$6.8 \times 10^5 \text{ pCi} \times 0.5 \text{ air changes/hour} = 3.4 \times 10^5 \text{ pCi/hour}.$$

The exponential decay time of radon is much longer than the half-hour air exchange time, so equation 2-1 simplifies to

$$\text{flow in} = \text{flow out, air exchange}$$

$$= 3.4 \times 10^5 \text{ pCi/hour}.$$

This result is in rough agreement with the estimates of various sources of radon entry into homes found by Brill et al. 1994 and shown in table 2-5.

Table 2-5 Contribution of Radon from Various Sources

Source	Estimated contribution (pCi / hour)
Soil gas transport	$0-5.4 \times 10^5$
Release from portable water	$0-2.2 \times 10^5$
Soil gas diffusion	$1.1 \times 10^4 - 2.2 \times 10^4$
Diffusion from building materials	$1.1 \times 10^3 - 1.1 \times 10^5$
Sum of midpoints	4.6×10^5

Solution 2-2b

The first step in this problem is to use the result from part (*a*) to find the radon concentration of 16 pCi/l. Since the ratio of concentration to flow is constant,

$$\text{flow at 16 pCi/l} = 16/1.5 \times 3.4 \times 10^5 \text{ pCi/hr} = 3.7 \times 10^6 \text{ pCi/hr}.$$

Now, imagine that we increase the airflow into the house so that the concentration becomes 1.5 pCi/l.

The model of airflow and radon concentration we developed above still applies; only the numbers have changed. First, note that radon removal via the decay route is considerably lower than the needed exit rate. Consequently, ignoring that route will not affect our calculation.

Let the ventilation rate be represented by the variable R (in units of ft^3 air/hour):

$$R \times 1.5 \text{ pCi/l} \times 28.3 \text{ l/ft}^3 = 3.6 \times 10^6 \text{ pCi/hour},$$

$$R = \frac{3.4 \times 10^6 \text{ pCi/hour}}{1.5 \text{ pCi/l} \cdot 28.3 \text{ l/ft}^3},$$

$$= 8.5 \times 10^4 \text{ ft}^3 \text{ air/hour}.$$

Divide by the volume of air in the house to convert the units of the answer to air changes per hour:

$$R = 8.5 \times 10^4 \text{ ft}^3 \text{ air/hour}/(16,000 \text{ ft}^3/\text{air change})$$

$$= 5.3 \text{ ACH}.$$

Note that we could easily have calculated this result from part (*a*) by considering how airflow, radon concentration, and radon flow are related. First, if airflow is constant, then radon flow is *directly proportional* to the concentration of radon in the air:

$$\text{radon flow} = \text{airflow} \times \text{radon concentration}.$$

The concentration of radon inside the part (b) house (16 pCi/l) is roughly a factor of ten (also referred to as an *order of magnitude*) greater than that in part (a), 1.5 pCi/l. If the airflow stays the same, then using equation 2-1, the radon flow should also increase by approximately a factor of ten, and it does (from 0.34×10^6 pCi/hour to 3.6×10^6 pCi/hour). The second part of the problem asks how much additional airflow is necessary to reduce the radon concentration in the house to 1.5 pCi/l. By rearranging the terms in the model, we see that if radon flow is constant, airflow is *inversely proportional* to radon concentration:

$$\text{radon concentration} = \frac{\text{radon flow}}{\text{airflow}}.$$

If we want to reduce the radon concentration by a factor of ten, then we have to increase the airflow by a factor of ten. Or, to be more precise,

$$\frac{C_2}{C_1} = \frac{R_1}{R_2}$$

so

$$R_1 = R_2 \times \frac{16}{1.5}.$$

Solution 2-2c

To answer this question, we need to introduce some basic thermodynamics, which, much like modeling, is a much less imposing concept than it might seem to the uninitiated. Specifically, it refers to stocks and flows of energy (and heat is a form of energy), as opposed to stocks and flows of matter. Homeowners are charged for the amount of energy that they use to heat their houses, regardless of what kind of furnace they have. We need to build a model relating the amount of energy necessary for heating the air in a house to the house's ventilation rate.

The goal of heating houses in the winter is to keep the air inside the house warmer than the air outside. Energy must be consumed continuously in order to maintain this temperature difference because heat escapes from the house in many ways. Heat is conducted outside through the walls and the windows, cold air enters the house through cracks and openings as warm air leaves, and the house radiates energy in the form of infrared radiation. The mechanism that we are concerned with here is energy transfer by the entry of cold air. The air that enters the house from the outside must be heated to the same temperature as the inside air in order for the temperature of the house to remain constant. Because changing the ventilation rate probably won't affect the other mechanisms of heat loss, let us group the effect of all of the others together in a single constant, P_0, which will designate the rate of energy consumption necessary to heat the house if the airflow rate is zero. The energy necessary to heat a volume V of cold air is

$$E = \rho \cdot V \cdot c \cdot (T_{in} - T_{out}), \qquad 2\text{-}2$$

where ρ is the density of air (the mass of air per unit volume), c is the specific heat of air (the amount of energy needed to raise the temperature of a specific mass of air by one temperature unit), T_{in} is the temperature inside the house, and T_{out} is the temperature outside. If cold air enters at a rate R, then the rate at which energy must be consumed to heat the flow to the required temperature is given by

$$P_1 = \rho \cdot R \cdot c \cdot (T_{in} - T_{out}).$$

The total rate of energy consumption necessary to heat the house is then

$$P = P_0 + P_1 = P_0 + \rho \cdot R \cdot c \cdot (T_{in} - T_{out}).$$

Since the problem states that air exchange with the outside accounts for 20% of heat loss, then

$$P_{orig} = 4\, P_{1,\,orig} + P_{1,\,orig}$$

$$= 5 \cdot \rho \cdot R \cdot c \cdot (T_{in} - T_{out}). \qquad 2\text{-}3$$

From this, we can determine that

$$P_{new} = P_{orig} + 10 \cdot R \cdot c \cdot (T_{in} - T_{out})$$
$$= 4 \, P_{1,orig} + 10 \, P_{1,orig}$$
$$= 14 \, P_{1,orig}$$

At this point, we could look up the density and specific heat of air for the temperatures we are considering and enter these numbers into the equation, but is that necessary? From equation 2-3, it is clear that the rate of energy consumption is proportional to one-fifth the ventilation rate. So, if the ventilation rate increases by a factor of ten, then the rate of energy use will approximately triple. In a month's time, the home-owner will have used approximately twice as much energy because of the ventilation increase, so the heating bill for that month will have increased by a factor of two! An analysis of the other options available to the homeowner would be necessary to evaluate this course of action, but based solely on the above estimate, and some knowledge of the cost of home heating energy, ventilation seems likely to be a very expensive solution.

Solution 2-2d

Ventilation works on one part of the overall stock-flow situation by increasing the flow out. Alternatively (and frequently in practice), many radon mitigation strategies focus on reducing the rate of radon flow in from soil and water by erecting physical barriers (sealing cracks, using water filters) and by reducing the pressure difference between the house and the surrounding soil. Typically, and especially during the heating season, the house is at a lower pressure than outdoors, as hot-air is less dense and therefore tries to rise (like a hot-air balloon). This forces air out the top of the house and draws it in toward the bottom. Hence air is also drawn up to some extent from the soil into the house in the cases of basements and slabs on grade. By decreasing the pressure in the soil outside the house (subslab ventilation) or increasing the pressure inside the house, the radon entry rate can be reduced. Crawl spaces between the living area and the ground can also be ventilated in lieu of ventilating the entire house, since radon concentration in the house due

Table 2-6 Radon Mitigation Strategies: A Comparison of Features

Technique	Typical radon reductions	Typical range of installation costs[a] (contractor)	Typical operating cost range for fan electricity and heated/cooled air loss (annual)[a]	Comments
Subslab suction (subslab depressurization)	80%–99%	$800–2500	$75–175	Works best if air can move easily in the material under the floor slab
Drain-tile suction	90%–99%	$800–1700	$74–175	Works best if drain tiles form complete loop around the house
Block-wall suction	50%–99%	$1500–3000	$150–300	Only in houses with hollow block walls; requires sealing job of major openings
Sump hole suction	90%–99%	$800–2500	$100–225	Works best if air can move easily to sump under slab or if drain tiles form complete loop
Submembrane depressurization in crawlspace	80%–99%	$800–2500	$50–175	Less heat loss than natural ventilation in cold winter climates
Natural ventilation in crawlspace	0%–50%	$200–500 if additional vents are installed; $0 if no additional vents	May be some energy penalties	Costs are variable

Sealing of radon entry routes	0%–50%	$100–2000	None	Normally used in combination with other techniques; requires proper materials and careful installation
House (basement) pressurization	50%–99%	$500–1500	$150–500	Works best with tight basement that can be isolated from outdoors and upper floors
Natural ventilation[b]	Variable	$200–500 if additional vents are installed; $0 if no additional vents	$100–700	Significant heat and conditioned air loss; operating cost dependent upon utility rates and amount of ventilation
Heat recovery ventilation	25%–50% if used for full house; 25%–75% if used for basement	$1200–2500	$75–500 for continuous operation	Limited use; works best in a tight house and when used for basement; less conditioned air loss than natural ventilation

[a]The costs provided in this exhibit represent the range of typical costs for reducing radon levels in homes above 4 pCi/l down to radon levels below 4 pCi/l; in most cases homes are reduced to an average of about 2 pCi/l; After Brill et al. 1994.
[b]This is the option presented in problem 2-2b.

to convective transport from the crawl space will be proportional to the crawl space concentration. Table 2-6 (from Brill et al. 1994) lists a number of these options.

Stock-flow models give us insight into two essential elements of the applied dose: how it arrived at and departs from the affected site, and how much is at that site. The dynamics of these systems can be defined with just a few parameters: flows in and flows out, and stock size.

Problem 2-2 shows how household radon exposure estimates are based on stock-flow models. Flows include both physical flows of radon from the ground and water sources, and out through ventilation, and loss through the decomposition of radon into radioactive daughters, which are also of concern. Such models help explain why houses built on slabs will have higher indoor radon concentrations than will those with subfloor spaces, and can also inform remediation efforts.

_____

Problem 2-A. Problem 2-2 Revisited

Repeat 2-2a assuming

    a. 0 ACH, 3.8 day half-life

    b. 1 ACH, 3.8 day half-life

    c. 0.5 ACH, no radon decay

Discuss your results. Which exit flow is more important? Why do you think this is? Why isn't the sum of (a) and (c) the same as your original answer?

_____

Problem 2-B. Equilibrium Concentration

    a. If the flow rate of radon into a 20,000 ft^3 house is 15×10^8 pCi/hr, and the ventilation rate is 3 ACH, calculate the equilibrium concentration in the house. (Assume that air within the house is evenly mixed.)

    b. Discuss the use and credibility of the "even mixing" constraint.

Physiologically based pharmacokinetic (PBPK) models are stock and flow models used to think about how chemicals move and concentrate within the body. While many dose-response experiments look only at the dose applied to the animal (in food, on air, on skin), understanding what goes on inside the body can provide much better information on why a given response occurs.

Problem 2-3. Simple PBPK Model—Continuous Dose

Assume a group of rats are fed 0.05 mg/day of chemical X for a long period of time. After several weeks, the level of X in the rat bodies reaches equilibrium, with 10% going through the digestive system without being absorbed and the rest absorbed by the blood, removed by the kidneys, and evacuated in urine. The total amount of X in a rat's body (not including that in the digestive system) is known to be 0.60 mg. The concentration of X in the kidneys is four times that in the blood. Assume that a rat has 500 ml of blood, and each rat kidney weighs about 5 g (see box 2-1).

a. Draw a line and box diagram showing the stocks and flows of X.
b. How much X is absorbed each day by the blood? How much passes through the digestive system without being absorbed by the body? How much is evacuated in urine?

Box 2-1. Weight of rat kidneys

In creating this problem, we were uncertain about the weight of a rat kidney, so we asked a colleague whom we thought would know. In turn, she asked another colleague, who provided an answer that examplifies the approach we're trying to teach. It follows verbatim:

Dear Ann:

Jeez, I dunno. The last rat kidney I saw was about 2 cm long by 1 cm fat, if I remember correctly... but the rat was kinda small. Assuming a prolate spheroid, that gives a volume of $\pi/3 = 1.05$ cubic cm. Assuming that flesh is about the same density as water, that's about one gram. So to an order of magnitude, the answer is between 1 and 10.

Sincerely, Henry

c. What is the concentration of X in blood? In the kidney?

d. Repeat (a)–(c), this time including the liver. Assume that the liver (also 5 g) has ten times as high a concentration of X as does the blood, and that half of the removal of X from the body is via breakdown in the liver of X into daughter chemicals.

Many removal mechanisms work as a function (linear or other) of concentration. The "biotransfer factor" (BTF) is a coefficient that indicates the concentration of a chemical in a tissue as a function of dose level, and is a constant if the removal process is a linear function of concentration. The BTF is the ratio of the amount of substance per amount of body weight to the rate of ingestion of the substance $[(\text{mg}_{substance}/\text{kg}_{tissue})/(\text{mg}_{substance}/\text{day})]$. It is used to calculate the concentration using the equation

$$\text{concentration (in the body or organ)} = \text{BTF} \times \text{dose}. \qquad 2\text{-}4$$

e. Calculate the BTFs for X in the blood, the kidneys, and the liver. Calculate the equilibrium concentrations if the administered dose is 0.01 mg/day.

f. Suppose that the mechanism by which the liver removes X is "saturable," such that it can metabolize only 0.07 mg per day, with a maximum liver concentration of 1.5 mg_X/kg_l. What will be the equilibrium concentrations of the blood, kidneys, and liver at a dose rate of 0.20 mg_X/day? Be explicit about any assumptions you make when solving this problem.

Solution 2-3

Solution 2-3a

Figure 2-5 shows the stocks and flows of X.

Solution 2-3b

If 10% of X passes through the system without being absorbed, the 90% is absorbed, so 0.005 mg/day pass through, while 0.045 mg/day

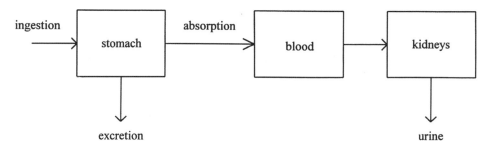

Figure 2-5. Line-box diagram depicting the flow of chemical X through the body, as described in problem 2-3a.

are absorbed. At equilibrium, the amount entering the body must equal that leaving, so 0.045 mg/l must be evacuated in urine.

Solution 2-3c

The body contains 0.6 mg X at equilibrium, and it is distributed between the blood and the kidneys. Assuming that blood and water have approximately the same density (1 mg/ml), 50 ml of blood is equal to 50 mg of blood.

The following equation describes the stocks of X, and their relationships:

$$Q_{total} = Q_{blood} + Q_{kidney} = C_{blood} \times M_{blood} + C_{kidney} \times M_{kidney}, \quad 2\text{-}5$$

where Q is quantity of X, C is concentration, and M is the mass of tissue or organ. Furthermore, it is given that $C_k = 4 \times C_b$, so

$$Q_t = C_b \times M_b + 4 \times C_b \times M_k.$$

Now there is one equation and one unknown. Rearranging,

$$C_b = Q_t/(M_b + 4 \times M_k),$$

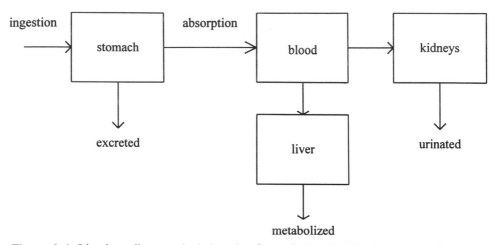

Figure 2-6. Line-box diagram depicting the flow of chemical X through the body, as described in problem 2-3d.

and plugging in the numbers,

$$C_b = 0.60 \text{ mg}/(0.5 \text{ kg} + 4 \times 0.005 \text{ kg/kidney} \times 2 \text{ kidneys})$$
$$= 1.1 \text{ mg}_X/\text{kg}_b,$$
$$C_k = 4 \times C_b = 4.4 \text{ mg}_X/\text{kg}_k.$$

Solution 2-3d

The same amounts of X are absorbed by blood and passed through. However, if half of the removal is not through urine, then the amount removed by the kidneys is half as much as before, or 0.023 mg/day (figure 2-6).

Next, add the liver to equation 2-5:

$$Q_t = Q_b + Q_k + Q_l = C_b \times M_b + C_k \times M_k + C_l \times M_l, \qquad \text{2-6}$$

so

$$C_k = 4 \times C_b,$$

and so

$$C_l = 10 \times C_b,$$
$$Q_t = C_b \times M_b + 4 \times C_b \times M_k + 10 \times C_b \times M_l.$$

Now there is one equation and one unknown. Rearranging,

$$C_b = Q_t/(M_b + 4 \times M_k + 10 \times M_l),$$

and plugging in the numbers,

$$C_b = 0.60 \text{ mg}/(0.5 \text{ kg} + 4 \times 0.005 \text{ kg} \times 2 + 10 \times 0.005 \text{ kg})$$
$$= 1.0 \text{ mg}_X/\text{kg}_b,$$
$$C_k = 4 \times C_b = 4.0 \text{ mg}_X/\text{kg}_k,$$
$$C_l = 10 \times C_b = 10 \text{ mg}_X/\text{kg}_l.$$

Solution 2-3e

If $C = \text{BTF} \times \text{dose}$, then $\text{BTF} = C/\text{dose}$. So the BTFs for X are

$$\text{BTF}_b = 1.0/0.05 = 20 \ (\text{mg}_X/\text{kg}_b)/(\text{mg}_X/\text{day}),$$
$$\text{BTF}_k = 4.0/0.05 = 80 \ (\text{mg}_X/\text{kg}_k)/(\text{mg}_X/\text{day}),$$
$$\text{BTF}_l = 10/0.05 = 200 \ (\text{mg}_X/\text{kg}_l)/(\text{mg}_X/\text{day}).$$

Note that, so long as the "passthrough" rate of 10% is independent of dose level, it does not matter whether the BTF is for administered dose or absorbed dose. If, on the other hand, it is dose dependent, then the equation for the BTF must either reflect this dependence or be based on absorbed dose.

If the administered dose is 0.01 mg/day,

$$C_b = \text{BTF}_b \times \text{dose} = 20 \ (\text{mg}_X/\text{kg}_b)/(\text{mg}_X/\text{day}) \times 0.01 \text{ mg}_X/\text{day}$$
$$= 0.2 \text{ mg}_X/\text{kg}_b,$$
$$C_k = \text{BTF}_k \times \text{dose} = 80 \ (\text{mg}_X/\text{kg}_k)/(\text{mg}_X/\text{day}) \times 0.01 \text{ mg}_X/\text{day}$$
$$= 0.8 \text{ mg}_X/\text{kg}_k,$$
$$C_l = \text{BTF}_b \times \text{dose} = 200 \ (\text{mg}_X/\text{kg}_l)/(\text{mg}_X/\text{day}) \times 0.01 \text{ mg}_X/\text{day}$$
$$= 2.0 \text{ mg}_X/\text{kg}_l.$$

Solution 2-3f

At 0.2 mg/day dose, the blood absorbs 0.18 mg/day. The liver cannot process half of this since it is limited to 0.07 mg/day. What is neither given nor clear without further information is whether the BTF for blood is dependent on or independent of the liver removal process. If it is independent, then the BTFs for blood and kidneys are known, and the equilibrium concentrations may be easily calculated:

$$C_b = \text{BTF}_b \times \text{dose} = 20 \ (\text{mg}_X/\text{kg}_b)/(\text{mg}_X/\text{day}) \times 0.2 \ \text{mg}_X/\text{day}$$
$$= 4.0 \ \text{mg}_X/\text{kg}_b,$$
$$C_k = \text{BTF}_k \times \text{dose} = 80 \ (\text{mg}_X/\text{kg}_k)/(\text{mg}_X/\text{day}) \times 0.2 \ \text{mg}_X/\text{day}$$
$$= 16 \ \text{mg}_X/\text{kg}_k,$$

and

$$C_l = \text{saturation concentration} = 1.5 \ \text{mg}_X/\text{kg}_l.$$

This works if the mechanism of removal is a linear function of body or organ concentration. If, on the other hand, the BTF for blood is dependent upon the liver removal process, then the equilibrium blood (and consequently kidney) concentration could be much higher.

Problem 2-C. Alternative Depictions

Provide alternative graphical depictions of problems 2-2 through 2-4. Explain why your pictures make the problem more or less clear.

Problem 2-4. PBPK—Finite Dose of Barium

While the last problem considered a constant dose, this one looks at a limited-time dose.

Barium (Ba) begins to have acute toxic effects on humans after 0.2–0.5 g has been ingested at one time, and a fatal dose is about 1 gram (Harte et al. 1991). The effects are due to Ba blocking the flow of potassium across cell membranes, thereby inhibiting their function.

Assume the following:

i. Available barium is transferred from the stomach into the blood at a rate that is "instantaneous" relative to the elimination rate

ii. The toxic effects are a function of blood Ba concentration, such that the effect at 0.2 g ingested is due to the resulting concentration of 0.04 g/liter of blood (an adult human body contains about 5.0 liters of blood)

iii. The body eliminates barium via the kidneys at a rate proportional to the concentration, such that

$$d[\text{Ba concentration}]/d_t = k \times [\text{Ba concentration}], \qquad 2\text{-}7$$

where $k = -0.007/\text{minute} = -0.4/\text{hour}$

a. Calculate the blood concentration associated with a fatal dose.

b. If someone swallows 0.8 g of barium, how long does it take for the concentration in the blood to drop to a subacute level (0.04 g/liter)? What is the concentration after one hour?

c. Suppose a worker is unaware that he has swallowed a chunk of a material that contains a harmless salt and 1.3 g Ba. Because it is trapped in the salt, the Ba is not immediately available to the bloodstream. Instead, it dissolves at a constant rate over a ten-hour period. Estimate whether the worker will experience acute toxic effects, and whether the dose is likely to be fatal.

d. Repeat (c), this time using 0.07 g and 2.4 g.

Solution 2-4

Solution 2-4a

The answer to this is a straightforward calculation. Using the assumptions that the barium is absorbed instantaneously and that the body contains 5 liters of blood, the "fatal" concentration is

$$C_{\text{fatal}} = 1.0 \text{ g Ba}/5 \text{ liters blood} = 0.2 \text{ g Ba/liter}.$$

Solution 2-4b

This problem differs from problem 2-3 in two important ways: The flow in is finite, rather than continuous, and the removal rate is dependent upon the concentration. As in problem 2-3, it may be useful to make a

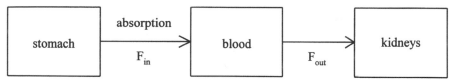

Figure 2-7. Line-box diagram depicting the flow of barium through the body, as described in problem 2-4.

simple diagram of the process (figure 2-7), and to recognize that both F_{in} and F_{out} are functions of the concentration.

The concentration of barium in the blood at any given time, $C(t)$, is determined by the initial concentration (C_0), the total that has entered the blood since that time, and the amount that has been removed since that time. Thus we can represent the concentration at any given time t as

$$dC(t)/dt = \text{flow in} - \text{flow out}. \qquad 2\text{-}8$$

The concentration at any given time T is the sum of the initial amount plus the amount that has entered the body since that time less the amount that has left, all divided by the volume of blood.

In this case, $C_0 = 0.16$ grams/liter if we assume that the 0.8 grams of barium is absorbed instantaneously. After the initial dose, there is no flow in, which leaves us with the equations

$$dC(t)/dt = \text{flow out} = -kC(t). \qquad 2\text{-}9$$

Next, let's take a look at equation 2-9. It says that, unless additional Ba is added to the blood, the change in concentration from one time to the next $(d[\text{Ba concentration}]/dt)$ is going to be negative, since $k < 0$ and concentration has to be ≥ 0. In addition, a higher concentration leads to a higher removal rate and consequently a more rapid reduction in concentration. So if the concentration is 0.16 g/l, the removal rate will be 0.0011 g/min, at 0.16 g/l will be 0.00011 g/min, and so on.

Rather than dive in to the calculus that allows us to solve this problem, it is informative to conceptualize the process. The last sentence above suggests a way to think about it. We know that, in this case, the initial concentration is 0.16 g/liter, and that after a long period of time, the concentration will approach 0. This implies a trend like that

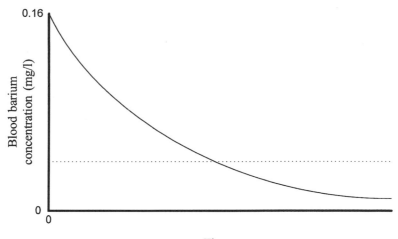

Figure 2-8. Expected long-term model form for blood barium concentration, given the conditions described in problem 2-4b.

shown in figure 2-8. What we don't know is how long it will take to get to, say, 0.04 g/liter (the "subcritical" concentration).

Looking at change over an increment of time, one can produce a spreadsheet approximation. One approach would be to estimate the removal amount for an initial time, apply this to some period of time, then do the same for the subsequent time period. If the resulting estimate is too rough, a smaller time frame can be used, until a sufficiently close approximation is reached, which implies that the approximate concentration at time t_1 can be found from

$$C(t_1) = C(t_0) - 0.007 \times C(t_0) \times (t_1 - t_0). \qquad \text{2-10}$$

As a first cut, consider 60-minute increments. Equation 2-10 then yields 0.16 g/l − 60 min × 0.007 × 0.16 g/l/min = 0.09 g/l after one hour. For the second 60-minute increment, $0.094 - 60 \times (0.007 \times 0.094) = 0.055$ after 2 hours, 0.032 after 3 hours, and so on. Table 2-7 shows the one-hour values using this method for 120, 60, 30, and 1 minute intervals.

What can we tell from this? Clearly, all estimates of this type will underestimate the concentration at any given time, but the smaller the increment, the more precise the solution. Since 0.040 is the concentration of concern, it is interesting to see that the two hour approximation

Table 2-7 Concentration at One Hour for Various Time Intervals

Time (hour)	Two hour	Concentration by increment		
		One hour	Half hour	One minute
0	0.16	0.16	0.16	0.16
1		0.093	0.10	0.11
2	0.026	0.054	0.062	0.069
3		0.031	0.039	0.045
4	0.0041	0.018	0.024	0.030
5		0.011	0.015	0.019
6	0.00066	0.0061	0.0095	0.013
7		0.0035	0.0059	0.0084
8	0.00011	0.0020	0.0037	0.0055

predicts under two hours to this level, the one- and one-half-hour approximations predict under three hours, while the one-minute approximation predicts over three hours.

The concentration after one hour is at least 0.11 g/liter.

While this approach provides an approximate model without using calculus, knowing a bit of calculus provides a quicker analytic answer.

Recall equation 2-9, $dC(t)/dt =$ flow out $= -kC(t)$.

From this, we can write

$$C(T) = \int_0^T [C_0 - kC(t)]dt.$$

We know that the concentration "drains away" at the rate of $-k(t)$, so the solution will be exponential. The integral or derivative of an exponential is another exponential, so in our case we have that if $dy(x)/dx = ky(x)$, then

$$y(X) = \int_0^X [y_0 + ky(x)]dx = y_0 e^{-kX}, \qquad \text{2-11}$$

which is an exponential decay equation. Substituting $C(t)$ and T gives

$$C(T) = \int_0^T [C_0 + kC(t)]dt$$
$$= C_0 e^{-kT}.$$

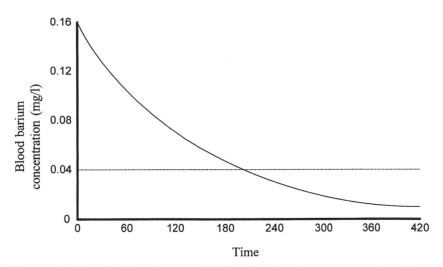

Figure 2-9. Specified long-term model for blood barium concentration, given the conditions described in problem 2-4b. 0.04 g/l is the "critical" point at which the effect is subacute. This level is reached after 200 minutes.

To solve for time, recall that if $y = e^x$, $\ln(x) = y$. Therefore, if $C(T) = C_0 e^{-kT}$, $T = -[\ln(C(T)/C_0)]/k$, and solving for $C(T) - 0.04$,

$$T = -\frac{\ln(0.04/0.16)}{0.007/\text{min}} = 200 \text{ minutes.}$$

Solving for concentration when $T =$ one hour (60 minutes),

$$C(60) = 0.16 \text{ g/l} \times e^{-0.007/\text{min} \cdot 60 \text{ min}}$$

$$= 0.11 \text{ g/l.}$$

From now on, simply recognizing that a constant fractional change per unit time always yields an exponential means we can write

$$C(t) = C(0)e^{-(0.007/\text{min})t}, \qquad \text{2-12}$$

and solve for t associated with the C of interest.

Figure 2-9 is a graph of this process. Note the shape of the curve, which characterizes the exponential decay function.

Solution 2-4c

This is similar to (b), except that now, rather than having an instantaneous dose, there is a constant flow in over a 10-hour period, and the initial concentration is 0. If we let flow in $= F_i$ (in grams/liter/minute), this can be represented as

$$dC(t)/dt = \text{flow in} - \text{flow out}$$
$$= F_i - kC(t)$$

and

$$C(T) = \int_0^T [F_i(t) - kC(t)]dt.$$

This is similar to equation 2-11, and the solution can be found in a basic calculus text under "first-order linear differential equations":

If

$$dy(x)/dx = a - ky(x),$$

then

$$y(X) = \int_0^X [y_0 + a - ky(x)]dx = \frac{a}{k} + \left(y_0 - \frac{a}{k}\right)e^{-kX}.$$

This is a modified exponential decay equation. Substituting F_i, $C(t)$, and $C_0 = 0$,

$$C(t) = \int_0^t [F_i + kC(t)]dt$$

$$\times \frac{F_i}{l} - \frac{F_i}{k}e^{-kt}.$$

Substituting the values,

$$C(t) = \frac{0.00044 \text{ g}/1/\text{min}}{0.007/\text{min}} - \frac{0.00044 \text{ g}/1/\text{min}}{0.007/\text{min}}e^{-0.007t}.$$

In this case, the concentration is 0 at time 0, and increases at a decreasing rate until time = 10 hours, at which point the source, or inflow, is "turned off," and the concentration begins dropping toward 0. Consequently, the maximum concentration will be reached at $t = 10$ hours. Thus,

$$C(10 \text{ hours}) = \frac{0.00044 \text{ g/l/min}}{0.007/\text{min}}$$

$$- \frac{0.00044 \text{ g/l/min}}{0.007/\text{min}} e^{-(0.007/\text{min} \cdot 600 \text{ min})}$$

$$= 0.062 \text{ g/l}.$$

Clearly, an acute concentration will be reached.

Solution 2-4d

We have already "solved" this problem by building the model above; all we need do now is crank the numbers.

For 0.07 g, $F_i = 0.00023$, so

$$C(10 \text{ hours}) = \frac{0.00023 \text{ g/l/min}}{0.007/\text{min}} - \frac{0.00023 \text{ g/l/min}}{0.007/\text{min}} e^{-0.007 \cdot 600}$$

$$= 0.032 \text{ g/l},$$

which is subacute.

For 2.4 g, $F_i = 0.008$, so

$$C(10 \text{ hours}) = \frac{0.0016 \text{ g/l/min}}{0.007/\text{min}} - \frac{0.0016 \text{ g/l/min}}{0.007/\text{min}} e^{-0.007 \cdot 600},$$

$$= 0.23 \text{ g/l},$$

which is in the fatal range. Figure 2-10 shows the results.

Problem 2-D. How Much Resolution Is Too Much?

Use the technique of solution 2-4a to calculate 10-second intervals. Does your answer suggest why we stopped with 1-minute intervals? Explain.

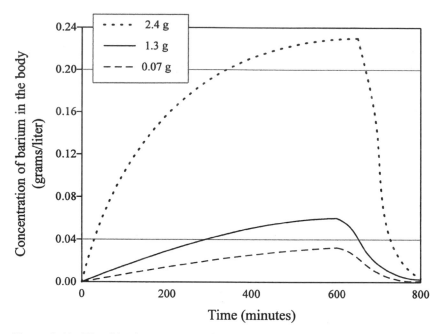

Figure 2-10. Blood barium concentrations over time corresponding to 0.07, 1.3, and 2.4 g initial intake.

Problem 2-E. How Much Information Is Needed?

Suppose you did not know how much blood a human body contained, but had all the other information provided. Could you still solve problem 2-4? How? What does this suggest about the dose information? Do you think it is equally applicable to a healthy adult and an already sick infant?

Problem 2-F. Sensitivity Analysis

Suppose you are uncertain about a number of the values: lethal concentration level, dissolution rate, blood Ba removal rate, and blood volume. In separate calculations vary each of these by 10%. Do all of the values influence the end result equally? Next, vary the same parameters by factors of two and ten. Discuss what these calculations suggest about where more precise information would be useful.

Cause and Effect Relationships

While stock-flow models provide information about exposure, cause-effect models help us conceptualize what happens given the exposure. For health issues, cause-effect models are referred to as dose-response models, but the concepts and mathematics developed extensively for those can be applied to many other relationships. This section starts with a simple linear dose-response relationship and then explores more complex relationships using a tried and true statistical tool: gambling. Once the more complex models have been developed conceptually, you are then asked to apply them to real situations.

Problem 2-5. Radon and Cancer

Problem 2-2 introduced exposure to radon; this problem looks at the possible effects of that exposure. Based on lung cancer deaths among uranium miners, the risk estimate for lung cancer deaths from exposure to radon decay products is perhaps 200–600 lung cancer deaths (LCD) per million working-level months [200–600 LCD/(10^6 WLM)].

The definition of a working-level month is exposure to the decay products in equilibrium with 100 picocuries[1] of radon per liter (pCi/l) of air for 173 hours a month (3.7 disintegrations per second = 100 pCi/l).

The average radon level in U.S. homes is about 1.5 pCi/l. It is ordinarily assumed in making risk estimates that people spend 75% of their time at home.

    a. How many working-level months of radon decay product exposure is the average person in the United States subject to each year? (Note that due to removal processes the decay products are not generally in equilibrium with their parent, radon, in indoor air. It

---

[1]Note that the standard international (SI) unit for radioactivity is the becquerel (Bq), equaling one decay per second. One Ci = 3.7×10^{10} Bq. In the interest of rationalizing units, the Bq was adopted around 1980, but U.S. agencies have generally continued to use curies.

is generally reasonable to assume that an "equilibrium" factor of 0.5 must be used in calculating decay product exposure in indoor air.)

b. Assuming that the linear no-threshold hypothesis (recall figure 1-2) is valid, what are your chances of contracting fatal lung cancer over a seventy-year lifetime if the living area of your home has a year-round average radon level of 1.5 pCi/l, 4 pCi/l, and 400 pCi/l?

c. On the same basis, assuming that the average U.S. exposure is 1.5 pCi/l, how many lung-cancer deaths would you predict in the United States each year as a result of the exposure of the population to radon progeny?

d. "Relative risk" (RR) is defined as the ratio of risk rate in an exposed population to that in the population as a whole from all causes. If the rate of lung cancer from all causes is 4%, what is the relative risk associated with living in a 400 pCi/l home?

e. What could be done to reduce this risk?

Solution 2-5

This problem uses the simplest dose-response relationship: linear. While the linear model may provide only a preliminary description of the risky phenomenon in question, it should give at least a ballpark indication of the possible outcomes. Subsequent chapters will address the assumptions used here in greater detail. Certainly, for a first "back of the envelope" calculation it is the right place to begin. More complex models will often not give very different results over a range of doses.

The reason for using the linear model for radon cancer is extensive research suggesting that the mechanism of radiation-induced cancer may require only one hit.

Solution 2-5a

Exposure can be calculated from the amount of time spent in the home during the year, the average concentration of radon in the home, the equilibrium factor, and conversion factors. Thus

Exposure

= fraction of time spent in home × concentration

× equilibrium factor × conversions

$$= \left(\frac{0.75 \text{ time home}}{\text{total time}} \right) \cdot \left(\frac{1.5 \text{ pCi}}{\text{liter}} \right) \cdot \left(\frac{8640 \text{ hr}}{\text{yr}} \right)$$

$$\cdot \left[\frac{1 \text{ WLM}}{(173 \text{ hr/mo}) \cdot (100 \text{ pCi/l})} \right] \cdot 0.5$$

$$= 0.28 \text{ (WLM/yr)}.$$

Solution 2-5b

Over a 70-year lifetime, total exposure in working-level months in a home with an average radon level of 1.5 pCi/l is 0.28 WLM/year × 70 years = 20 WLM

The range of lung cancer death probabilities at this concentration is

$$\text{low: } (20 \text{ WLM}) \cdot \left(\frac{200 \text{ LCD}}{10^6 \text{ WLM}} \right) = 0.0040 \text{ LCD},$$

or 0.40% chance of LCD and

$$\text{high: } (20 \text{ WLM}) \left(\frac{600 \text{ LCD}}{10^6 \text{ WLM}} \right) = 0.012 \text{ LCD},$$

or 1.2% chance of LCD.

Since we assume that the effect is linear, the LCDs for 4 pCi/l and 400 pCi/l can be calculated as

$$\text{LCD}(4 \text{ pCi/l}) = (4/1.5) \times \text{LCD}(1.5 \text{ pCi/l}) = \{1.1\% \text{ to } 3.2\%\},$$

$$\text{LCD}(400 \text{ pCi/l}) = (400/4) \times \text{LCD}(4 \text{ pCi/l}) = \{110\% \text{ to } 320\%\}.$$

Solution 2-5c

Using a U.S. population estimate of 270×10^6 people, the predicted range of lung cancer deaths per year in the United States is

$$\text{low: } \left(\frac{0.28 \text{ WLM}}{\text{person} \cdot \text{yr}} \right) \cdot \left(\frac{200 \text{ LCD}}{10^6 \text{ WLM}} \right) \cdot 270 \times 10^6 \text{ people}$$

$$= 15{,}000 \text{ cancer deaths per year,}$$

$$\text{high: } \left(\frac{0.28 \text{ WLM}}{\text{person} \cdot \text{yr}} \right) \cdot \left(\frac{600 \text{ LCD}}{10^6 \text{ WLM}} \right) \cdot 270 \times 10^6 \text{ people}$$

$$= 45{,}000 \text{ cancer deaths per year.}$$

Solution 2-5d

$$RR = \text{exposed/baseline} = \{110/4 \text{ to } 330/4 = \{28 \text{ to } 83\}.$$

The risk for those in 400 pCi/l homes relative to those in 1.5 pCi/l homes is $400/1.5 = 270$.

Solution 2-5e

To reduce the risk, people could:

- Spend more time outdoors (note that the 75% factor affects exposure)
- Improve ventilation in residence and office buildings and homes
- Seal cracks and other possible leaks from below into the home

Federal and state governments could

- Engage in programs of diagnosing radon levels in people's homes, and identify high-risk areas (note: in the next chapter we introduce distributions of radon in different locations, which improve our ability to decide where this type of intervention is most viable)
- Improve models for estimating home radon concentrations, exposures, and effects

Problems 2-1 and 2-5 examined the radon issue only superficially, but even these rudimentary models provide an idea of the extent of the possible risks in the United States from household exposures. Variability and uncertainty in exposure levels will be the subject of the next chapter.

Mechanistic Models and Curve Fitting

Problem 2-5 assumed that the relationship between exposed population and cancer risk had a linear curve. However, as we have tried to emphasize, the shape of the curve implies something about either the data used to find it or a belief in the underlying mechanism causing it. It may be the case that the relationship between radon exposure and lung cancer risk is linear, but then again it may not. The following section explores some possible mechanisms that might cause certain curve shapes, as well as some techniques for simply fitting a curve to data, regardless of possible mechanisms. Either modeling or simple curve fitting can be used (with caution) to extrapolate to exposures beyond the data range.

Extrapolation from a limited data set is frequently necessary to find quantitative risk estimates. Just as assumptions about mechanism have a profound effect on curve shape, the shape of the curve can have a profound influence on the extrapolated values. There are two reasons to select one curve over another: theoretical and empirical. Ideally, both will be available, but at other times, judgment must be used to fill in the gaps. The following series of problems provide a conceptual understanding of the reasoning behind some basic theoretical and empirical models.

Problem 2-6. Conceiving "Mechanistic" Models

Consider the following four games. How might each help us depict some risk process?

Game 1: Flip a Fair Coin

In this game, a gambler wins if she gets a head on a single flip of a fair coin.

Game 2: Consecutive Flips of a Fail Coin

In this game, a win comes as soon as two heads come up on repeated throws of a coin.

Game 3: Roll a Die and Draw a Card

In this game, a win is achieved upon a die roll of six and the drawing of ace of hearts from a deck. In the first period, the player rolls a die. If a six shows, the player can draw a card the second period; if not, the player must roll the die again. Once a six has been rolled, the player gets to draw a card each subsequent period until the ace of hearts turns up. Assume that the dealer reshuffles the whole deck after each draw.

Game 4: Roll a Die and Draw a Card, with Reversibility

This is identical to game 3, except that, if the ace of hearts is not drawn within three periods after rolling a six, the player must begin rolling the die again.

Solution 2-6

Much of statistics was developed to assist wealthy gamblers to better understand games of chance. It is interesting to find that a few examples from games of chance can allow us to conceptualize a variety of possible risk-related processes. The models considered below are called mechanistic, because they are meant to describe mathematically the deterministic underlying mechanisms of the risk-laden process. The mathematical derivations of these models are explored more completely in chapter 5, and they will be applied to a range of problems; for now, we consider them conceptually.

Game 1

If asked to model the number of wins expected from repeated flips, one would (hopefully) come up with a model that looked very much like figure 1-2, with a slope of $1/2$. Here, the "dose" is the flip of the coin, and the "response" is the number of heads. In risk parlance, this is a "one-hit" model, since for any given dose (flip), a single head wins the game. Russian roulette, where a single bullet is placed in one of six chambers of a revolver, and the barrel is then spun and the gun fired, represents a graphic example of how a one-hit (or in this case one-shot) model might apply to risk assessment. The calculation can be quite simple—the probability of a hit is 1 in 6, and the consequence is death. Unfortunately, this example sweeps under the rug a ubiquitous concept, uncertainty (e.g., due to the occurrence of misfires). Uncertainty (which will be the focus of chapter 4) is fairly easy to ignore in this case, since the possibility of a misfire may reduce both the probability and the consequence by a bit, but not to a significant extent.

Game 2

This describes a simple "two-hit" model. An example of a multihit process is driving a nail, where six consecutive strikes (or three for a good carpenter!) will result in the desired outcome, while fewer than that will not. This example also illustrates a further complication, heterogeneity. It takes one person six hits and another three to achieve the same effect.

Game 3

This represents a multistage process, which differs from the multihit in that a first stage must be completed before the second stage can begin. The first stage is the die roll, which has a one in six chance of occurring each period. Once the player is in the second stage, she has one chance to win each period by drawing the ace of hearts. A simple application of this game is a mechanical production unit with a backup. The main system may have a one in a hundred chance of jamming each operation, after which it reverts to the backup, which has a one in ten failure rate each period.

The multihit and multistage models can be combined as appropriate, to describe increasingly complex processes. In the production example above, the main unit might continue to function after one jam, but fail upon a second jam. Here the first stage requires two hits, and the second stage only one.

Game 4

This game represents another level of complexity for the multistage or multihit model: the possibility of reversibility. In the production process example, there may be someone who comes by every ten operations to check for jams. While this modification improves the model as a descriptive tool, it also becomes increasingly difficult to describe with an equation. While a person might be able to decide quickly how much to bet in each turn of game 1, betting preference for game 4 is not so transparent. Add to this uncertainty and heterogeneity (what happens if the worker gets distracted? if the night-shift worker is more qualified than the one on swings? if the probability of a jam is actually one in eighty?), and the models become quite murky. Nonetheless, they may still be better than doing no analysis at all.

Applications of Mechanistic Models to Real Risks: Examples

Why are we interested in being able to describe these kinds of mechanisms? Because we may have theoretical or empirical reasons to believe that they fit a particular process of concern. Models ranging from one-hit to multistage/multihit as well as those with reversibility can represent many processes observed in the world.

Carcinogenesis

The mechanisms of carcinogenesis have been the subject of decades of research. In 1972, Knudson and Strong suggested that a two-hit model is appropriate for certain cancers, where two discrete and independent cellular defects are required for a tumor to begin. The heterogeneity issue was salient in Knudson's research, since his model was based on the finding that certain individuals are born with one of the hits, an inheritable genetic deficiency (Weinberg 1996).

Birth Control

A major public health issue in the United States today is the use of condoms to avoid Acquired Immunne Deficiency Syndrome (AIDS). A multistage model for the risk might include a first stage, the decision whether or not to use the condom, a second where there is some probability that the condom will fail, and a third, where there is some probability of contracting AIDS given condom failure. If this is the right model, useful conclusions can be drawn about whether it is more efficacious to invest in education on condom use, improving condom reliability, encouraging abstention, or researching a vaccine. Beyond this, of course, are the moral implications: what role should the government play in decisions about sexually transmitted disease? At times in these moral debates, experts and science may be used as gladiators, sent in by different sides to fight a battle that is in actuality one of divergent preferences, not science.

Investments

A final example of a multihit system is the investment portfolio. Investment consultants consistently urge their clients to diversify, in effect rendering their financial risk multihit, since a broad-based economic collapse is necessary to deplete the individual's capital. Add to this insurance on some of those investment, and failure becomes a multihit multistage process. Federal Deposit Insurance Corporation (FDIC) insurance may represent a "threshold" situation for investors who have some money in protected bank accounts. While failure of the entire FDIC system is possible, for most purposes it is sufficient to guarantee that failure of the institution will have no effect (ignoring inconvenience) on the investor up to a point.

Problem 2-7. Using the Wrong Model, Getting the Model Wrong

a. Discuss the ramifications of placing a bet when you think you're playing game 3 in problem 2-6 (roll a die, pick a card), but you're actually playing game 1 (flip a fair coin).
b. Suppose that you were playing game 1 (flip a fair coin), but for some reason you calculated that the chance of getting heads was only one in four rather than 50/50. How would this affect your decisions?

Solution 2-7

Solution 2-7a

The mechanistic models described by the games above are based on theories about how processes in the world work. However, theories can be wrong, and the empirical evidence may be insufficient to choose the right model. The implications of choosing the wrong model can often be profound. Consider the following examples of common-mode failures.

Redundancy is very common in safety systems. Ideally, redundancy means that failures are either multistage, multihit, or a combination, and the probability of system failure is calculated based on this model. Unfortunately, if this model turns out to be incorrect, the overall risk may be much greater. Common failure modes, where a single event can knock out both (or all) components of a supposedly redundant system, represent the misapplication of the multihit model to a one-hit reality. Such was the case when the control cable system for a nuclear power plant design called for "redundancy" by having many "redundant" cables, rather than a single one. Since the cables were bundled, the system was in fact no better off in a catastrophic failure (the cables caught on fire) than had it been a single cable! (Schneider 1991).

A similar example was the "backup" electrical wiring and battery system on some early DC-3 airplanes. In the event of a major short-circuit or electrical fire, the backup system was engaged. The problem, however, was that the pathway of the backup systems followed the primary pathway very closely. In several cases, even small fires destroying one system knocked them both out simultaneously.

One might consider the Challenger disaster as a one-hit model. If each flight is considered a "dose," there was a "hit" on the seventy-eighth event. There was no redundancy, or second stage. The ramifications of this model are only apparent in retrospect for the one case, but may have lessons for us in the future. Are there similar shuttle systems that might need only one hit to result in failure? If so, what is the best option—to decrease the probability of that hit coming about, or to add a second stage to the failure sequence? This question is explored more thoroughly in problem 8-2.

Alternatively, the failure might be represented as a two-stage system, where the dose is the planned launch, the first stage is the decision to launch, and the second stage is the launch itself. One argument suggests the failure to abort resulted when a number of individuals found themselves under pressure to approve a launch at the decision stage, whereas some of them might have objected had they been anonymous. If this is the case, then the first stage might be modeled as an actual one-hit stage, the group decision, that was meant to be a multihit stage, where any individual in the group could have called for the flight to be canceled. Many airlines take the approach that any ground crew member can hold up a takeoff, in effect requiring a multihit sequence for a system failure.

Solution 2-7b

In this case, you might bet considerably more than fifty cents for the chance of winning a dollar if the coin comes up tails. In the long run, you could lose a lot of money this way. Clearly, this is an artificial problem, since it is very unlikely that you would get so basic a calculation wrong. However, consider something more complex, such as game 4: it might be very easy to make an unnoticed mistake in the calculations. Consequently, even when the right model of a process is found, great care must be taken to insure that it is derived correctly.

The Approximate Two-Stage Model. Versions of the two-stage model have been used for predicting low-dose carcinogenesis for years and, as discussed above, the model has many other useful applications. However, a recent paper demonstrated some ways in which the approximate derivation of the equation most people used for the model was glaringly incorrect under certain circumstances (Cox 1995). If the first stage is much more improbable than the second, the second stage dominates the approximate model. The first stage limits the process, both in reality and in the exact solution. Similarly, the approximation only works for convex to linear processes, and any concave process would appear linear. There is no evidence that the problems with the approximate model have led to incorrect regulatory decisions, but it is easy to see how they could (Chiu et al. 1999 explore this issue). It is also important to note that the mathematics involved in the exact solution are not terribly complicated; it had simply been assumed that the approximation was "good enough."

Problem 2-8. Empirically Derived Dose Response

Consider another game. An activity has been done 1000 times. The player is given a complete sequential list of the outcomes of those 1000 tries. The outcomes of those 1000 events are summarized as 510 *H*s and 490 *T*s, and statistical tests show no consistent pattern explaining the likelihood of an *H* or a *T* for a given event. This process will be repeated, and the game is won when an *H* is the outcome. How would you bet on this game?

Solution 2-8

This game should be easy to bet on, since it looks so much like the expected value of game 1. The outcome could be modeled as a one-hit model, but this would be based exclusively on empirical data, not on theoretical understanding of the process. Modeling empirical findings can be very useful, so long as important parameters are not ignored. An example is the actuarial tables used by automobile insurance companies, founded primarily on data for drivers as a function of age, sex, location, and past history.

Decisions about preferring empirical or theoretical models become more problematic when data are incomplete and extrapolation is necessary. An excellent example of this is the ongoing debate over basing the human carcinogenicity potential of a potential toxin on high-dose animal toxicity tests. A major question is whether the processes that lead to observed tumors at high doses are equivalent to those for low doses, or if there is some fundamental difference so that you cannot simply "scale" the results from high to low dose. Suggestions for dealing with this range from finding a "best fit line" for the data (model-free curve fitting) to using a precise mechanistic model.

Given the quality and nature of the arguments for various approaches, the EPA has recently proposed dropping the "linearized multistage" it has used since 1978, in favor of an extrapolation methodology similar to the that used by the Food and Drug Administration (FDA) (U.S. EPA 1996). Rather than assume there is an all-purpose model, these agencies require a linear extrapolation from the lowest measured dose, with exceptions made on a chemical-by-chemical basis, as better information is available. This is a "science policy" decision, and while problematic in scientific terms, such decisions are unavoidable when actions must be taken absent full information.

Putting Stock-Flow and Dose-Response Models Together to Explore Issues of Food Safety, Climate Change Models, and Earthquake Damages

The FDA has recently undertaken a marked and highly publicized shift in its approach to minimizing contamination of meat. In the past, meats were evaluated at the end of the line, where they were inspected primarily for gross contamination. Thus food safety was reduced to a one-hit process, evaluating a single stock at a given time, while ignoring flows. In sharp contrast, the new system considers flows—requiring microscopic analysis at various points in the butchering process. Now, flows as well as stocks of contaminants are tracked, and a contaminated carcass must pass through many stages before it is released. This multistage, flow-based evaluation can anticipate the eventual stocks even better than the old method allowed with direct observation!

Climate change models are enormously complex. However, stepping back from those models and looking at the broad processes, it is clear that the two basic modeling elements dominate. Even the enormously complicated global climate models (GCMs) are essentially stock-flow models, looking at how much carbon there is in the earth, in the air, and in the oceans, and how it moves from one of these reservoirs to others. Cause-effect models provide estimates of climate change as a function of these carbon stocks and flows. In this case, internal feedback can be an important factor, as dose rates are a function of existing levels. Some of the more troubling implications are the possibilities of positive feedback. Both empirical data and theoretical constructs are used, and uncertainty is ubiquitous; nonetheless, understanding of the process begins by looking at the major elements.

Problem 2-9. Earthquakes versus Traffic Risks

The death toll in the 1989 Loma Prieta earthquake in California's Santa Cruz Mountains was approximately 60. Traffic on the San Francisco/Oakland Bay Bridge stopped for two months. Consequently, ridership on BART (Bay Area Rapid Transit) increased from 200,000/day to 300,000/day, implying that at least 100,000 fewer commuters per day drove across the Bay Bridge for 60 days before the Bay Bridge was repaired.

a. Explicitly identify the relevant stocks, flows, doses, and responses, and some of the key uncertainties involved.

b. Estimate the number of traffic deaths averted.

c. Address the following question: was the earthquake lifesaving or life-taking on average? Is this a reasonable way to assess the impact of the earthquake?

d. Empirical verification: how many deaths actually occur each day in the Oakland to San Francisco commute? (Answer not given—this can be a research exercise.)

To solve this problem, you may need to use the following assumptions (as well as to make some of your own):

  i. An average commuting distance of 30 miles (each way)

  ii. About 80 million cars in regular use in the United States

  iii. 10,000 miles/year/car

  iv. 30,000 traffic deaths in the United States per year.

  v. 50,000 deaths in the United States per year due to air pollution, perhaps half from automobiles

  vi. Negligible number of deaths on BART each day

Solution 2-9

Solution 2-9a

The stocks and flows are the commuters either in cars or by BART. The main stock of interest is the number of people on the roads at commute time, since it is these people who are exposed to the risk of accidents.

There are two dose responses of interest. The first is given with one datum: number of people killed by the Santa Cruz earthquake. If this number were not known, it could be approximated by looking at similar earthquakes (magnitude and location). The Kobe, Japan, Los Angeles, California, and Mexico City earthquakes might be reasonable comparisons.

The other dose response is accidents per vehicle mile traveled. Ideally, this would be location specific—perhaps data on commuting deaths in the Bay area. Instead, this problem uses the relatively crude, but probably order-of-magnitude accurate, deaths per year in the United States. The pollution deaths are highly uncertain, both the number estimated and the contribution of automobiles. In addition, it might be relevant to consider that it was autumn when commuter traffic was reduced: smog is less of a problem in the San Francisco Bay area in the autumn through spring than it is during the summer. On the other hand, commuter traffic produces a disproportionate amount of pollution because it is often stop and go.

Solution 2-9b

Include the following additional assumptions:

Average ratio of people to cars in the United States: 1.0 people/car
Average ratio of commuters to cars in the San Francisco area: 1.2
commuters/car
Total number of deaths/year in the United States from motor vehicle
use: 55,000

(30,000 from accidents plus 25,000 from pollution).

Then,

$$\left(\frac{55000 \text{ deaths}}{\text{year U.S.}}\right)\left(\frac{\text{U.S.}}{260 \times 10^6 \text{ people}}\right)\left(\frac{1 \text{ person}}{\text{car}}\right)\left(\frac{\text{car} \cdot \text{year}}{10000 \text{ miles}}\right)$$

$$= 2.1 \times 10^{-8} \frac{\text{deaths}}{\text{mile}},$$

$$\left(\frac{30 \text{ miles}}{\text{car} \cdot \text{commute}}\right)\left(\frac{2 \text{ commutes}}{\text{day}}\right)(60 \text{ days})(10^5 \text{ commuters})$$

$$\times \left(\frac{1 \text{ car}}{1.2 \text{ commuters}}\right)$$

$$= 3.0 \times 10^8 \text{ miles},$$

$$(3.0 \times 10^8 \text{ miles})\left(8.5 \times 10^{-8} \frac{\text{deaths}}{\text{mile}}\right)$$

$$\approx 26 \text{ deaths averted.}$$

(Note: elsewhere in this book we use other figures for population of the
United States. 260 million was appropriate for 1989.)

Solution 2-9c

The earthquake, on average, was life-taking, killing three times more
people than the average of motor-vehicle-related deaths over the period
that the bridges and roads were under repair. Note that there is some

ambiguity here, since most of the earthquake deaths were traffic related. Nonetheless, catastrophic events such as earthquakes are seen to
be qualitatively different from the "routine" risk of driving a car.

This comparison of life loss from everyday commuter motor vehicle
traffic and the effect of the earthquake does not take into account
significant questions. First, there are economic losses that occur due to
an earthquake from the destruction of private property, infrastructure,
and harder-to-enumerate lifestyle disruptions. At the same time, several
new construction projects may be required, which will create economic
opportunities. Some assumptions may have limited validity—there is
the distinct possibility that there is a different number of commuters
per vehicle than the assumption; also, all vehicles are not cars. This
distinction may have effects on the number of accidents and the amount
of pollution.

Note also that the study, in comparing loss of life, does not take into
account non fatal injuries—from either automobile accidents or earthquakes.

Driving an automobile is a risky endeavor, but it is often considered
"voluntary." In contrast, even though people make a conscious decision
to live in earthquake-prone California, that risk is generally considered
to be involuntary. The assumption that driving is a voluntary risk may
require us to discount some of the damage caused by commuting. An
evaluation of this "discounting" may be done, in part, through an
economic examination of risk behavior under uncertainty. Nonetheless,
even if the number of traffic deaths was greater than the sixty lost in the
earthquake, it is not at all clear how (or even whether) to compare these
numbers. Certainly, the general sentiment of our society is that sixty
deaths from an earthquake are far more tragic than the same number
from automobiles or smoking.

Alternative Analysis

Perhaps we must look not only at commuters but at all members of the
San Francisco community. Problem 2-9a can be answered differently
below.

total traffic deaths = 51,500/year, or 8583 for a two-month period.

Of these,

$$\text{collision deaths} = 6315,$$
$$\text{non-collision deaths} = 2267.$$

The San Francisco area population in 1990 was 723,959 people. As a percentage,

$$\frac{723,959 \text{ people in S.F.}}{250 \text{ million U.S. pop.}} = 0.3\% \text{ of U.S. population in S.F.}$$

Taking 0.3% of all U.S. traffic deaths in a two-month period, we get

$$6315 \times 0.003 \approx 20 \text{ deaths.}$$

Each year, there are an estimated 50,000 deaths due to air pollution, about half of which are associated with automobile exhaust. That means that in a two-month period, there are about

$$(50,000 \text{ deaths/year}) \times (1/6 \text{ year}) = 8,333 \text{ deaths}$$

in the United States due to pollution, and

$$8,333 \text{ deaths}/2 = 4167 \text{ deaths}$$

due to pollution associated with automobiles.

Again, if about 0.3% of the U.S. popultion is in the San Francisco area, this means

$$4167 \text{ deaths} \times 0.003 = 12.5 \approx 13 \text{ deaths}$$

due to San Francisco area automobile pollution. Thus, total deaths averted = 19 + 13 = 32 < 60 deaths from the earthquake. By this calculation as well, it was a net *life-taking* earthquake.

Conclusion

With the introduction of risk assessment into new fields, as well as its increasing salience in those where it has long been a tool, it behooves a democratic society to use transparent methods, evaluated from multiple viewpoints. If taken on a case-by-case basis, risk assessment appears enormously complex, to the point of dissuading many from entering a particular fracas. If, on the other hand, interested parties understand the general methods, discourse about the details—especially value-laden assumptions—will improve.

References

Brill, A. B., Becker, D. V., Donahoe, K., Goldsmith, S. J., Greenspan, B., Kase, K., Royal, H., Silberstein, E. B., Webster, E. W. (1994). "Radon update: facts concerning environmental radon levels, mitigation strategies, dosimetry, effects and guidelines." SNM Committee on Radiobiological Effects of Ionizing Radiation. *Journal of Nuclear Medicine* 35(2):368–85.

Chiu, W. A., Hassenzahl, D. M., and Kammen, D. M. (1999). "A comparison of regulatory implications of traditional and exact two-stage dose-response models." *Risk Analysis* 19(1):27–35.

Cox, L. A., Jr. (1995). "An exact analysis of the multistage model explaining dose response concavity," *Risk Analysis* 15(3):359–68.

Crump, K. (1996). "Risk of benzene-induced leukemia predicted from the pliofilm cohort." *Environmental Health Perspectives* 104(6):1437–1441.

Crump, K. S. and Allen, B. C. (1984). *Quantitative Estimates of the Risk of Leukemia from Occupational Exposures to Benzene. Final Report to the Occupational Safety and Health Administration.* Ruston, LA: Science Research Systems.

Harte, J., Holdren, C., Schneider, R., and Shirley, C. (1991). *Toxics A to Z.* Berkeley, CA: University of California Press.

Knudson, A. G., Jr, and Strong, L. C. (1972). "Mutation and cancer: neuroblastoma and pheochromocytoma." *American Journal of Human Genetics*, 24(5):514–32.

Price, P., Nero, A. V., and Revzan, K. (1998). *U.S. Radon Map from the Lawrence Berkeley Laboratory High-Radon Program.* Berleley, CA: Lawrence Berkeley National Laboratory.

Schneider, U. (1991). "Introduction to fire safety in nuclear-power-plants." *Nuclear Engineering And Design* 125(3):289–95.

Wallace, L. A., Pellizzari, E., Leaderer, B., Zelon, H., and Sheldon, L. (1987). "Emissions of volatile organic compounds from building materials and consumer products." *Atmospheric Environment* 21(2):285–393.

Weinberg, R. (1996). "How Cancer Arises." *Scientific American* (September):62–70.

U.S. EPA. (1996). "Proposed guidelines for carcinogen risk assessment," *Federal Register* 61(79) (April 23):17960–18011.

3

Review of Statistics for Risk Analysis

Introduction: Statistics and the Philosophy of Risk Assessment

Statistics is fundamental to quantitative risk analysis, and those readers who have not taken a basic statistics class are strongly encouraged to do so. For those who have had statistics, this chapter should be a review, presented in the context of several important risk issues. Even readers who have strong statistics backgrounds are encouraged to skim the chapter, as it introduces some risk concepts that will be useful in later chapters. Readers without much exposure to statistics can use this chapter as an introduction and tutorial in the basic set of statistical skills they will need. All of the problems in this book can be solved with the statistics covered in this chapter or within later text. This chapter is not, however, a substitute for a thorough grounding in statistical techniques and underlying theory.

The process of constructing and solving problems for risk assessment can be viewed as two broad types of activity, building a useful mental picture, or model, of the system, and getting reasonable values into the model. Each of these activities could be the subject of many volumes. The goal of this book is to develop a working knowledge of the fundamental risk assessment paradigm, using basic descriptive models solidly grounded in statistical analysis.

The previous two chapters introduced some risk concepts, but dealt with uncertainty only qualitatively. Statistics allows risk analysts to address at least part of the uncertainty quantitatively, by placing values on some of the known uncertainty and by describing some of the natural variability. It is important to keep in mind that the precision provided by statistical analysis is only as good as the numbers that go in, since in almost all risk contexts there is profound uncertainty that cannot be quantified. Nonetheless, evaluating those aspects that are known will always be better than ignoring them.

Statistics to many people looks like—and often, jokingly, is said to be—a manipulation of the *numbers* to determine what conclusions can and cannot be drawn from them. While this can be true at an operational level, statistics is really about the *models* being used. Data are virtually always incomplete, and for many of the most interesting problems they are terribly incomplete. The goal of statistical analysis is to decide what model provides a reasonable or "best" fit to the data, and then to use that description to learn more about the system.

This chapter begins with statistics that describe data—the mean and median—and then moves to a basic statistical model, the normal (or Gaussian) description of a data set. It also introduces some more complex models that will be useful in future chapters and which address the limitations of the simple ones. The development of a larger set of analytic and computational tools, however, does not warrant neglect of the basic tools set forth here; many people skip these steps and miss many revealing insights as a result! Working through the problems in this book by sequentially applying simple and then more sophisticated models should provide sufficient familiarity with each tool to permit efficient determination of when each will be useful in tackling increasingly complex cases.

This same "statistical" philosophy of moving from simpler to more involved descriptions of the data applies to much of the analysis that we will explore in this book. In fact, as unfamiliar as risk assessment is to many scientists, policy analysts, and policy makers, it often involves no more modeling than keeping track of the information on a problem in the sort of clear categories that statistical methods provide.

The second part of the risk assessment procedure involves getting the "right" numbers into the models. As complex as people sometimes make this out to be, it too often involves little more than understanding how much and what type of data are necessary to discriminate among competing models. As the problems become more complex, this task of course becomes increasingly difficult, although the models themselves often remain conceptually straightforward. Many scientists are fond of paraphrasing Einstein, who said that models should be as simple as possible, but no simpler. In this book we will attempt to demonstrate that approach in understanding the risks from environmental, technological, political, and economic systems and decisions.

Problem 3-1. Average Radon Exposure

Given the data in table 3-1, what is the average U.S. resident's yearly risk of getting cancer from household exposure to radon?

Solution 3-1

Solving this problem requires figuring out what parameters are relevant and modeling the relationship among those parameters. Once the model has been built, plugging in the numbers and calculating the result—in this case cancer death rate (CDR)—can be trivial.

- The model needed to solve this problem describes the cancer rate as a function of radon exposure. The parameters relevant here describe exposure to the radon gas (concentration and time period) and its health effect, or

$$CDR = f(\text{radon cancer potency, concentration, exposure time}).$$

Table 3-1 Data on Radon Exposure

Parameter	Value	Units	Source
Mean household radon concentration	1.5	Picocuries per liter (pCi/l)	Nero 1986
Time the average individual spends in the house	15–18	Hours per day	EPA default assumption (EPA 1992)
Radon cancer potency	1×10^{-8}	Cancer deaths per hour of exposure at 1 pCi/l	NAS-BEIR IV 1988

- While the exact relationship between these parameters is a matter of ongoing research, let us assume that your risk of dying from cancer increases linearly with the concentration to which you are exposed and the amount of time of the exposure that you receive:

$$CDR = \text{carcinogenic potency of radon}$$
$$\times \text{concentration} \times \text{exposure time}$$
$$\text{(usually multiplied by some conversion factors)}.$$

Once this simple risk model is built, numbers can be plugged in, and the risk calculated:

$$CDR = \left(\frac{10^{-8} \text{ deaths}}{(\text{pCi}/\text{l}) \cdot \text{hour}} \right)(1.5 \text{ pCi}/\text{l})\left(\frac{15 \text{ to } 18 \text{ hours}}{\text{day}} \right)$$
$$\times \left(\frac{365 \text{ days}}{\text{year}} \right)(1 \text{ year})$$

$$= 8.1 \text{ to } 9.8 \times 10^{-5} \text{ deaths/year/average residential exposure}$$

$$\text{to } 9.8 \times 10^{-5} \text{ deaths/year/average residential exposure}$$

Now that the model is posited, specified, and calculated, many questions remain. Perhaps the most important question is, "What can be learned from this average national cancer rate?" From the standpoint of the individual or the policy maker, the answer may be "not much."

Why? Because this average exposure value provides no information on the distribution of radon exposures. If almost all U.S. houses are near this level, the policy implications are very different than if 99% of houses have negligible exposure levels, but 1% have very high levels, and 0.001% have extremely high levels. This book will develop the tools needed to understand the accuracy, value, limitations, and shortcomings of this simple model. At the end of the book, this same question will be revisited; by that time, you should have the tools needed to identify what is needed to improve and use the radon data.

Problem 3-A. Radon Exposures in Different Regions

Given the following data on mean radon levels in area homes, calculate the radon exposure risk for residents of

a. Louisiana (0.28 pCi/l) (Nero et al. 1986)
b. Iowa (2.72 pCi/l) (Nero et al. 1986)
c. Reading Prong, Pennsylvania ($\approx$ 10 pCi/l) (after Brill 1996)

Problem 3-2. Working with Data

Table 3-2 shows hypothetical radon levels of all the houses in a very small town. What is the mean radon level? The median radon level? What is the standard deviation of radon levels?

Solution 3-2

The Mean

There are two ways to calculate the mean in this problem. The first way, and also the most common method in risk analysis, is useful when each measurement has equal weight. Radon level is called a *variable* because it can take different values at different homes. Adding up the measurements and dividing by the number of measurements gives the mean.

This calculation of the mean is derived from the following general method. Let μ be the mean, N the number of data points, and let X_i

Table 3-2 Home Radon Levels

Home	1	2	3	4	5	6	7	8	9	10	11
Radon level (pCi/l)	4.04	4.60	5.73	5.39	2.37	5.39	4.60	5.05	4.38	5.05	4.04

represent each data point in the population, $i = 1, 2, \ldots, N$. Then

$$\mu = \frac{1}{N} \sum_{i=1}^{N} X_i.$$

3-1

The mean radon level calculated this way is

$$\mu = \frac{1}{11} (4.04 + 4.60 + 5.73 + 5.39 + 2.37 + 5.39$$

$$+ 4.60 + 5.05 + 4.38 + 5.05 + 4.04) \text{ pCi/l}$$

$$= 4.60 \text{ pCi/l}.$$

The answer is *not* 4.604 or 4.6. The data in table 3.2 imply that the radon concentrations were measured to the hundredth of a curie; consequently, it is not credible to arrive at an answer with a term for thousandths of a picocurie, but the measurements are more accurate than 4.6 would imply. In this calculation there are three *significant figures*. If the radon level for house 9 had been 4.4, that for house 2 4.6, and that for house 11 4.0 there would be only two significant figures for that term, and it would be impossible to say whether the real value were 4.55, 4.64, or something in between. Thus the calculated value for the mean would change to 4.6 pCi/l, since the data themselves are insufficiently precise to allow a more precise determination.

A second approach to finding the mean is to weight each value by the fraction of the data that it comprises (the fraction of houses at each radon level). This would be useful if the data specified only the values and the frequency with which they appear, rather than the individual house values (table 3-3).

As in the first case there is a general method for calculating the mean using fractional probability. Let N be the number of possible states that the data can take (in this case, $N = 7$), x_i the states, $i = 1, 2, \ldots, N$ (in

Table 3-3 Fraction of Houses at Each Radon Level

Radon level (pCi/l)	4.04	4.60	5.73	5.39	2.37	5.05	4.38
Fraction of houses at that level	2/11 (0.18)	2/11 (0.18)	1/11 (0.09)	2/11 (0.18)	1/11 (0.09)	2/11 (0.18)	1/11 (0.09)

this case, $x_1 = 3.04$, $x_2 = 4.60, \ldots, x_7 = 4.38$), and p_i the probability of the ith condition. Then

$$\mu = \sum_{i=1}^{N} p_i X_i. \qquad\qquad 3\text{-}2$$

The mean radon level calculated in this way is

$$\mu = \left(\frac{2}{11} \times 4.04 + \frac{2}{11} \times 4.60 + \frac{1}{11} \times 5.73 + \frac{2}{11} \times 5.39 + \frac{1}{11} \right.$$

$$\left. \times 2.37 + \frac{2}{11} \times 5.05 + \frac{1}{11} \times 4.38 \right) \text{pCi}/\text{l}$$

$$= 4.60 \text{ pCi}/\text{l}.$$

The Median

The median is found by arranging the data in ascending (or descending) order, and selecting the middle data point (or, if there are an even number of data, averaging the two middle points). In our case, the data points (radon levels) in ascending order are shown in table 3-4

In this case, the median is identical to the mean, 4.60 pCi/l. There are advantages and disadvantages to using the mean rather than the median. A strong argument is simply tradition- -it's the mean that is generally used to measure the central tendency of a distribution. The mean is the "center of gravity" of the data values, with half of the *summed* value above and half below, where the median is the center in the sense that half of the individual numbers are above and half below. By definition, the mean minimizes the sum of squared differences between itself and each of the data points. It is a continuous and consequently differentiable function.

The median, which minimizes the sum of absolute values of the differences between itself and each of the data points, is discontinuous.

Table 3-4 Radon Concentrations in Ascending Order

House number	5	1	11	9	2	7[a]	8	10	6	4	3
Radon concentration		2.37	4.04	4.04	4.38	4.60	4.60	5.05	5.05	5.39	5.39	5.73

[a] This value (4.60) is the median.

One property of the median that can be useful is its resilience to outliers. While the addition of a measurement that is much larger than all others in a sample may have a large effect on the mean, it is unlikely to change the median much, if at all. This is especially relevant if you believe that the outliers may represent bad data (Wonnacott and Wonnacott 1990). Caution is warranted, though: if the data are good, it may be the outliers you are interested in. For example, the addition of a house with a reading of 40 pCi/l to table 3-2 could represent an erroneous reading or a relatively high-risk house. It is left to the reader to calculate and compare the mean and median with this additional point.

Problem 3-3. Mean and Median: Why Worry?

Find the mean and the median for the radon concentrations in table 3-5. Why is the median a problematic measure in this case?

Solution 3-3

The median in this case is 3.88 pCi/l, while the mean is 6.33 pCi/l. The former is just below the U.S. EPA's "level of concern" of 4 pCi/l, while the latter is above it. If the median, rather than the mean, were used as the decision criterion for the area, the outcome would be very different. Note that this issue does not usually occur, because the EPA has intended its guideline to apply to individual houses, not to areas.

The mean and the median capture only one aspect of the data, and say nothing about what may be a very important aspect of these data, clustering in a "high group" and a "low group." Since radon levels have a number of causes, it is possible that all of the houses in the high group have something in common (such as no basement, a particular kind of

Table 3-5 Alternate Set of Household Radon Concentrations

House	1	2	3	4	5	6	7	8	9	10	11
Radon level (pCi/l)	3.83	3.54	3.46	3.90	12.06	8.97	9.24	14.67	3.13	3.88	2.99

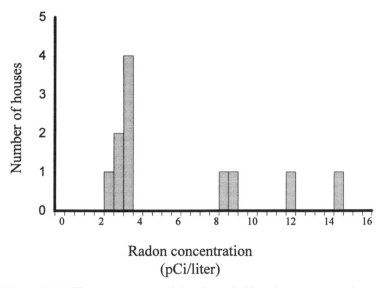

Figure 3-1. Histogram summarizing household radon concentrations from table 3-5.

bedrock, etc.), which is much more useful information than the mean or median alone. Figure 3-1 is a histogram of these values.

Measuring Dispersion

If you were to pick a random house from the data in table 3-2, you would "expect" the concentration to be the mean, or 4.60 pCi/l. But clearly this will not always be the case. The most common way to think about how well a mean represents the data in general is to think about how much a particular datum can be expected to vary from the mean. One statistical tool that is used to describe this is the *variance*, which is the average of the squared differences between each data point and the mean. More often, we refer to the *standard deviation*, which is the square root of the variance.

The standard deviation is calculated using the following general method. Let σ be the standard deviation (σ^2 is called the variance), N the number of data points, and let X_i represent each data point in the sample. Then

$$\sigma = \sqrt{\frac{1}{N} \sum_{i=1}^{n} (X_i - \mu)^2}\,.$$
  3-3

For our data set,

$$\sigma = \sqrt{\frac{1}{11}[(4.04 - 4.60)^2 + (4.60 - 4.60)^2 + (5.64 - 4.60)^2 \\ + \cdots + (4.04 - 4.60)^2]}$$

$$= 0.88 \text{ pCi/l}.$$

One reason for squaring the difference between each data point and the mean is that it removes the negative signs; otherwise, data points above the mean would cancel out with data points below the mean. If this were the only problem, then absolute values could be used instead of squares. However, squares are more tractable in a number of ways, including that they result in differentiable functions.

The Normal Distribution

What do we learn from the standard deviation? That depends largely on how the data are distributed. A frequent assumption in statistical analysis is that the values that a single variable can take follow a *normal* distribution, which is the well-known "bell-shaped" curve. While you should always question whether it is a reasonable assumption, when it is it can be very useful. An important property of the normal distribution is that 95% of all data will fall within 1.96 (or about 2) standard deviations on either side of the mean. Similarly, one standard deviation on each side covers about 70% of the data. Consequently, if we know the mean and the standard deviation, we can get a good idea of where most of the data are.

The following formula gives us a range (low value, high value) within which we would expect to find 95% of the data:

$$\text{range} = \{\bar{x} - 2\sigma, \bar{x} + 2\sigma\}. \qquad 3\text{-}4$$

For our data set,

$$\text{range} = (4.61 + 2 \times 0.88 \text{ pCi/l}, 4.61 - 2 \times 0.88 \text{ pCi/l})$$

$$= (6.37 \text{ pCi/l}, 2.84 \text{ pCi/l}).$$

Consider figure 3-2. The normal distribution is not obviously a good fit for the data in table 3-2. By "fit," we mean the extent to which the

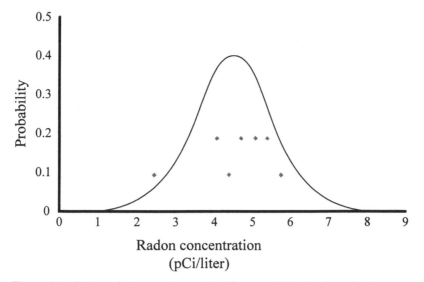

Figure 3-2. Data points superimposed with normal distributions for house-hold radon concentrations from table 3-5.

equations specified for the model (here the normal distribution with mean 4.61 and standard deviation 0.88) resemble the available data. If the data points fall on or very close to the points predicted by the model, there is a good fit. If the data points and the predictions are dissimilar, the fit is not good, and a better model should be sought. Later we will look into how to test such a fit statistically.

An important aspect of the distribution is that while 95% of the data are expected to lie in the (low value, high value) range, not every point within that interval is equally likely. The importance of the shape of the distribution will be explored later in the chapter, and used in examples throughout the remainder of the book.

_____

Problem 3-4. Sample Data Revisited

Suppose that, instead of representing all of the houses in a small town, table 3.2 is a random selection of houses from a very large town. What are the estimated mean radon level and standard deviation for the whole town? Within what range of radon levels do you expect 95% of all houses to fall? Solve this assuming that this is a random sample; then discuss why this might not be a good assumption, and the consequences if it is not.

Solution 3-4

Originally we assumed that the data in table 3-2 represented radon levels for all possible houses. In that case, it could be considered *population* data, since the 11 data points are the entire population of possible data points. The mean and standard deviations that we calculated above were "true" measures; we call them the population mean and standard deviation.

The new assumption here represents the more common case in which we have only a *sample* of possible measurements—perhaps all addresses were put in a hat, and eleven houses picked for testing, thereby generating a sample of measurements from the population of all measurements. We now use the sample data to approximate the true population mean, and we will want some idea of how closely it resembles the true mean, which we do not know.

Sample Mean

Let $\bar{x}$ be the sample mean, n the number of data points in the sample (in this case, $n = 11$), x_i each data point, $i = 1, 2, \ldots, n$ (in this case, $x_1 = 4.04$, $x_2 = 4.60, \ldots, x_{11} = 4.04$). Then

$$\bar{x} = \frac{1}{n} \sum_{i=1}^{n} x_i. \qquad \text{3-5}$$

In this case, $\bar{x} = 4.60$ pCi/l.

Sample Standard Deviation

Let s be the sample standard deviation (s^2 is the sample variance). The s is used in lieu of the σ to differentiate between the sample value calculated from the limited number of data that are available and the "true" value, which can only be known if all of the data are known. Let n be the number of data points in the sample and let x_i represent each data point in the sample. Then, similarly to equation 3-3,

$$s = \sqrt{\frac{1}{n-1} \sum_{i=1}^{n} (x_i - \bar{x})^2}. \qquad\qquad 3\text{-}6$$

It is important to note that in the case of the sample variance, $(n - 1)$ is used in the denominator rather than n. This is because we do not know the true mean μ but instead are approximating it with a statistic we generated, the sample mean ($\bar{x}$). The data points were determined independently; that is, given the first 10 data points, we could not have predicted what the 11th point would be. However, if we had the first 10 data points and the sample mean, we would know what the 11th data point would be. Consequently, by calculating the sample mean, we have given up one independent (free) data point. Since we use the sample mean to calculate the dispersion around that mean, we acknowledge that lost independence by giving up one degree of freedom. Calculating the sample standard deviation using $(n - 1)$ reflects this.

If the eleven data points above are only a sample of the population, not complete population data, the sample standard deviation is

$$s = \sqrt{\begin{array}{l}\frac{1}{10}[(4.04 - 4.60)^2 + (4.60 - 4.60)^2 + (5.73 - 4.60)^2 \\[1mm] + \cdots + (4.04 - 4.60)^2]\end{array}}$$

$$= 0.96 \text{ pCi}/\text{l}.$$

The added uncertainty associated with having only a sample implies a larger distribution of radon levels than was found in equation 3.3, where the value of $\sigma = 0.88$ pCi/l.

Why might we be concerned about this sample? This first reason is size: it is quite likely that 11 houses randomly selected from a large city will very poorly represent the city as a whole. Problems 3-5 and 4-3 will address this issue. Other problems arise when the sample is not truly random. For example, if the 11 houses in the sample were tested because the owners wanted the test, because they were easy to test, or because they were suspected of having high radon concentrations, then there may be *selection bias* in the sample, meaning that its mean will differ systematically from the true population mean. This bias can be up or down.

_____

Problem 3-5. Hypothesis Testing and Confidence Intervals

Once we have some information about some of the individual data, we would like to know how well the sample represents the entire set. The next problem explores how statistical tools describe the limits of our understanding. This includes not only the critical values discussed above (mean, median, etc.), but also confidence intervals, which describe the range of likely actual values based on the sample, and hypothesis testing, which allows us to formally test whether a value other than what we have calculated is likely or unlikely to be the actual value.

Table 3-6 shows the results of measurements from randomly selected houses in the same hypothetical town (including the 11 used above). Calculate the mean, median, and standard deviation for this new sample. Test the hypotheses that the actual (or "true") mean is equal to the mean (6.33 pCi/l) calculated in problem 3.3. Test the hypothesis that the true mean is equal to the EPA concern level of 4.0 pCi/l. How sure are you that the true mean radon level was under 6.00 pCi/l? Provide a 90% confidence interval for the calculated sample mean.

Solution 3-5

This problem requires a measure of the probability that the true mean value is accurately approximated by the small sample of data available.

Table 3-6 Larger Data Set (43 Homes) of Household Radon Concentrations

Number	1	2	3	4	5	6	7	8	9
Radon level (pCi/l)	4.04	4.60	5.73	5.39	2.37	5.39	4.60	5.05	4.38

Number	10	11	12	13	14	15	16	17	18
Radon level (pCi/l)	5.05	4.04	3.48	5.25	1.80	4.93	3.83	2.90	2.87

Number	19	20	21	22	23	24	25	26	27
Radon level (pCi/l)	6.48	3.74	3.99	0.89	3.72	3.51	4.47	1.72	3.26

Number	28	29	30	31	32	33	34	35	36
Radon level (pCi/l)	6.01	3.40	3.96	2.82	3.41	2.73	3.08	3.25	6.08

Number	37	38	39	40	41	42	43		
Radon level (pCi/l)	5.15	2.73	5.87	3.77	0.74	4.01	1.22		

As the number of measurements grows, so should confidence in the accuracy.

Mean, Median, Standard Deviation

Using the methods demonstrated above,

$$mean = 3.85 \ pCi/l,$$

$$median = 3.83 \ pCi/l,$$

$$standard \ deviation = 1.40 \ pCi/l.$$

Hypothesis Testing

The mirror image of the confidence interval is the *hypothesis test*. Strictly speaking, the data set defines only the mean of the sampled points, so there is some amount of statistical *uncertainty* about the "true" or population mean. We often want to test a hypothesis that the true mean is some number other than the estimate generated from the sample. For example, we might be concerned that the mean is above some threshold level of concern. Just as the confidence interval provides a range of certainty around the sample mean (expressed as a probability or percentage), the hypothesis test lets us reject a hypothesis, with some probabilistic or percent certainty of being right.

Two types of hypothesis testing that are often used are the *Student's t* test, which will be applied in this problem, and the *chi-squared* (χ^2) test, which will be explored in problem 3-7. (For tables of numerical values associated with these tests, as well as more information on how these equations are derived, see the appendixes.) Before looking at those two, it is useful to consider a more basic test, the *Z-score*, which is based on the standard normal, or Gaussian, distribution. If there are many data points, and we can assume their error terms are normally distributed, then the following equation can be used to test the hypothesis that the true mean is other than the sample mean:

$$Z = \frac{\bar{x} - \mu_0}{s/\sqrt{n}}.$$ 3-7a

Here, $\bar{x}$ is the sample mean, $s/\sqrt{n}$ is the sample standard error, and μ_0 is the hypothesized true mean. Alternatively, when we have proportions rather than values,

$$Z = \frac{\bar{x} - np}{\sqrt{p(1 - p)} / \sqrt{n}}.$$ 3-7b

A useful way to think about how to use this to test data is to consider the case where the sample mean is exactly equal to the hypothesized mean. In such a case, the denominator is irrelevant, since the numerator is zero. This suggests that the smaller the Z-score, the more likely the hypothesis is correct. If the sample mean and hypothesized mean are very different, the Z-score will be much larger. Looking at the standard deviation, it is evident that a very large s value will make the Z-score very small. This implies that if there is lots of variability in the data, it will be harder to reject the hypothesis that $\bar{x}$ and μ_0 are different. Finally, as sample size increases, the denominator decreases, and the Z-score increases. This implies that the more data there are, the more evident are the differences between $\bar{x}$ and μ_0.

To use the Z-score, plug in the calculated or known values for n, s, $\bar{x}$, and μ_0. Then refer to the Z-score tables in appendix A. If $Z > 1.94$ (≈ 2), the null hypothesis can be rejected at a 95% confidence level. Alternatively, the Z-score can be used to determine the confidence level at which the hypothesis may be rejected.

The central limit theorem says that with sufficiently large sample size, generally taken to be $n \geq \approx 30$, the variation of a sample mean will be normally distributed. This allows us to use the Z-score to compare the sample mean to the hypothesized mean, after they are "standardized" by dividing by the standard error. Once the Z-score is calculated, the consequent probability that $\mu_0 = \bar{x}$ can be looked up in a table such as the one in appendix A.

Student's-t

The Student's t-distribution is similar to the standard normal, except that it takes into account the limitations of small sample size. It has thicker "tails," that is, a larger number of data points or events located far from the mean. These tails incorporate the uncertainty that results from having sample rather than population data. Since the uncertainty goes down as the number of data points in the sample goes up, the tails of the t-distribution change with sample size as well.

The Student's t-statistic is calculated using the same formula as the Z-statistic:

$$t_{n-1} = \frac{\bar{x} - \mu_0}{s/\sqrt{n}}, \qquad \text{3-8}$$

and as $n \to 8$, $t \to$ normal distribution. The t-value can be looked up in appendix B. Because the t-statistic converges to the Z-value as sample size gets arbitrarily large, the t-value is often used as the default.

As sample size becomes very large, the Student's t converges on the Z. For values of n smaller than about 30, however, the distribution begins to look very different from the normal. Instead of finding 95% of the data within 1.96 standard deviations from the mean, less and less of the data are likely to be found in that interval as the sample size shrinks. Table 3-7 and figures 3-3a and b show the relationship between the t and the normal distribution as a function of sample size.

We can use equation 3-8 to test the hypothesis that the true mean equals 4.60 pCi/l as follows:

$$t_{42} = \left| \frac{3.85 - 4.60}{1.40/\sqrt{43}} \right| = 3.51.$$

Note that the absolute value is used. Because the t-distribution is symmetric and centered on zero, the probability of a value being below a negative value is equal to the probability that it is above the absolute value of that number. Consequently, t-tables usually give only positive values and assume this conversion will be made.

Referring to the t-table in appendix B for 42 degrees of freedom, if $t = 2.416$, $\alpha = 0.01$ (see figure 3-3). Because the calculated t-value is greater than $t^{\alpha=0.01}$, the hypothesis that $\mu = 4.60$ pCi/l (as found in problem 3-2) can be rejected at a $1 - \alpha = 1 - 0.01 = 99\%$ confidence level. Again, increasing t-values imply decreasing probability that $\bar{x} = \mu_0$.

Table 3-7 Comparison of Z and Student's t Statistics as a Function of Sample Size

n	1	2	3	5	10	15	25	30	40	50	100	∞
t	6.314	2.920	2.353	2.015	1.812	1.753	1.708	1.697	1.684	1.676	1.660	1.64
Z	1.64	1.64	1.64	1.64	1.64	1.64	1.64	1.64	1.64	1.64	1.64	1.64

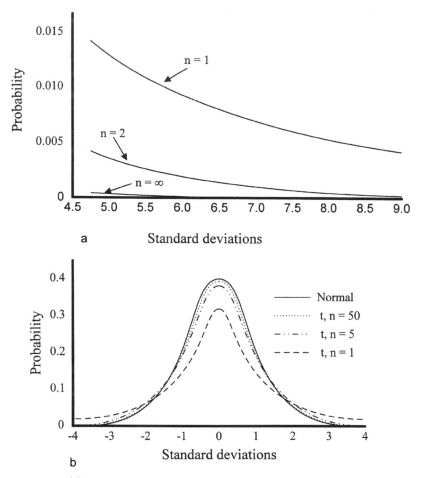

Figure 3-3. (a) The tails of the normal and *t*-distributions. Note that the increase from one to five degrees of freedom causes a significant reduction in the probability at the extremes. (b) Normal and *t*-distributions. Notice that the *t*-distribution with only one degree of freedom contains considerably less probability at the center, and considerably more at the extremes.

One-Sided and Two-Sided Tests

We sometimes want to know the range of possible values around the estimated value; at other times, we just want to be certain that one extreme will not occur. Often in risk analysis it is important that a risk be below a maximum level, which necessitates a *one-sided* test. Finding the probability that the mean radon level is below some maximum uses the same t formula that we used for confidence intervals and hypothesis testing. In this case, we calculate a t-value using the sample mean and the maximum level of concern (4.0 pCi/l) and find the confidence level associated with that t-value and the relevant degrees of freedom:

$$t = \left| \frac{3.85 - 4.00}{1.40/\sqrt{43}} \right| = 0.70.$$

Looking at the t-table for 42 degrees of freedom, we find that if $\alpha = 0.10$, $t = 1.3$. Because the calculated t-value is less than $t^{0.1}$, we cannot reject the hypothesis that $\mu = 4.0$ pCi/l at a 90% confidence level.

As shown in figure 3-4, the t-distribution is symmetric with respect to a vertical axis through the mean. Consequently, the α below a negative number is the same as the α above the absolute value of that number. From the t-table, $t_{42}(\alpha = 0.1) = 1.3$, so the hypothesis that the larger sample mean is equal to the EPA level of concern, 4 pCi/l, cannot be rejected.

A one-sided test is likewise used to test whether the mean is lower than 6.00 pCi/l:

$$t = \left| \frac{3.85 - 6.00}{1.40/\sqrt{43}} \right| = 10.1.$$

Since $t_{42}(\alpha = 0.01) = 2.42 < 10.1$, then we are confident at a 99% level that the true mean is less than 6.00 pCi/l.

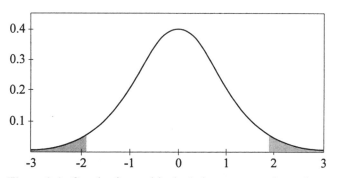

Figure 3-4. Graph of t_{42} with shaded α upper and equal α lower.

Confidence Intervals

The statistical tool that is used to determine how well we think the sample mean reflects the true mean is the confidence interval, or the probability that an acceptably wide interval around the sample mean $\bar{x}$ encompasses the true mean μ. Often a 95% confidence interval is used, but as we shall see later in this book, we may find it crucial to be much more than 95% confident in some numbers (such as the chance of an uncontrolled nuclear reaction). While a high confidence interval sounds good, a 5% or even 1% chance of being wrong is of course far too high for everything from airline crashes to contraceptive failure.

In the table in appendix B, α is the probability that the true mean is beyond (above or below) some critical value. In the case of the 90% confidence interval, we are concerned with limiting both the high and the low range, so $\alpha = (1 - .9)/2 = 0.05$. In other words, we tolerate a 5% possibility of overestimating and a 5% possibility of underestimating the real mean. Since we have $(n - 1)$, or 42, degrees of freedom, the critical value will be associated with a t-score, using equation 3-8 and appendix B, is $t_c = 1.68$, where t_c is called the "critical" t-score. Using this critical value, we can calculate the 90% confidence interval as follows:

$$\text{C.I.}_{.90} = \bar{x} \pm t_c s / \sqrt{n}$$

$$= [\,3.85 \pm 1.68 \times 1.4/\sqrt{43}\,]\,\text{pCi}/l$$

$$= \{3.49\ \text{pCi}/l,\ 4.21\ \text{pCi}/l\}. \qquad\qquad 3\text{-}9$$

In other words, there is a 90% certainty that the range {3.49 pCi/l, 4.21 pCi/l} bounds the true value.

Problem 3-6. Making Decisions

Can you be certain at the 95% level that the data in table 3-2 preclude the mean calculated for the data in table 3-6? At a 99% level? What possibility do we accept when we choose a certain level at which to accept or reject a null hypothesis?

Solution 3-6

We can test the hypothesis that the old mean is the new sample mean,

$$t_{10} = \left| \frac{4.60 - 3.85}{0.96/\sqrt{11}} \right| = 2.57.$$

Looking at the t-table for 10 degrees of freedom, we find that if $t = 2.76$, $\alpha = 0.01$, and if $t = 1.81$, $\alpha = 0.05$. Because the calculated t-value is greater than $t^{0.05}$ and less than $t^{0.01}$, we can reject the hypothesis that $\mu = 3.85$ pCi/l at the 95% but not the 99% confidence level. What does this tell us? One possibility is that there is a real difference between the original subset and the city as a whole. Another is that the difference lies in the small original sample size. The bigger the data set, the more robust the results; changing the value of one data point in the first subset would have a more significant effect than changing the same data point in the larger sample.

Type I and II Errors

The above problems have developed several methods for calculating and testing the accuracy or confidence of various estimates. However, there are fundamental ways that we can go wrong. If we reject a hypothesis that is in fact true, we have made a type I error. If we accept a hypothesis that is wrong, we have made a type II error. Type I errors can be simple mistakes, although they may be made for two complex reasons: (1) by selecting an inappropriate model or hypothesis, and (2) by putting too much faith in an insufficient or nonrepresentative data set. Deciding whether to avoid type II errors has important policy implications. For example, most atmospheric scientists agree that there has been a 0.4–0.6 degree centigrade increase in global temperatures over the past century. A type II error would occur were we to accept this as within natural climatic variability if the truth is that it is man-induced climate change.

Unfortunately, there is an inevitable tradeoff between the two types of errors. If we choose to be increasingly certain of not rejecting true hypotheses, the possibility that we will accept false hypotheses increases. A useful way to think about type II errors is to consider the philosophy favored by the U.S. criminal justice system, which prefers letting the guilty go free over punishing the innocent. In fact, "beyond a reasonable doubt" is often considered to be similar to a one-sided confidence interval, where "reasonable" is analogous to α. An important lesson from this is that, even in so "analytical" a tool as statistics, there are inherent value judgments being made about where to set confidence levels. As soon as a climatologist goes from stating that the climate change is likely to be due to anthropogenic sources to saying that society should accept that hypothesis, value choices have come into play. Remember that being right 95% of the time means being wrong one time in every twenty!

Distributions

Distributions are used to describe how frequently different values of a variable occur, and they allow us to evaluate such issues as the most likely value and the probability of a given value coming up. There are two general reasons to assign a particular distributional form to a model. The first is theoretical: the nature of the processes that give rise to the variable of interest may suggest one form over another. The second is empirical: the available observations of the variable may fit well to one distribution or another. In either case, the distributions can tell us something about the variable. If we theorize that it will follow one distribution, but in fact it does not, our theory may be incorrect (alternatively, the data may be inadequate). On the other hand, fitting the best available distribution to a data set may lead to further understanding if we have a general understanding of the types of behavior and phenomena that generate that distribution.

Problem 3-7. Moving Away from Ignorance

This problem demonstrates the use of the uniform and triangular distributions.

On average, one tornado has hit the (hypothetical) town of Barton Park in each of the past one hundred years, with no evident trend.

a. Draw a graph indicating the day that you expect a tornado to strike Barton Park in a one-tornado year. What is the probability that it will occur on July 4?

b. Assume that all of the past tornadoes have occurred in the months July through September, and that meteorological conditions are only right for tornadoes during that time, but no more information is available. Again, draw the distribution for a one-tornado year. What is the probability that a tornado will occur on July 4 now? January 10?

c. Assume instead that you suspect that a tornado will most likely occur on August 15, it is as likely to occur before then as after then, but it will not occur before July nor after September. Again, depict this graphically. What is the probability that the one tornado will occur on July 23?

Solution 3-7

Solution 3-7a

The simplest distribution, and thus often a first model of an unknown or poorly understood process, is the uniform distribution, where every possible outcome is equally likely. The chance of getting a particular value on the roll of a die is a good example. While the uniform is generally not a useful distributional form (since very few random events have multiple and equally likely outcomes), it will be useful later in the book for Bayesian analysis. Figure 3-5 depicts the uniform distribution for equally likely events in a given year (excluding leap years).

Since there are 365 days in the year, and the one expected tornado could occur on any of those days with equal likelihood, then the probability that it will occur on any given day is $1/365$, which is also the probability that it will occur on July 4.

Solution 3-7b

In this case, a little more information is given, but not much. The interpretation is that the probability is distributed equally across part of the total data range (three months out of the year). This is often referred to as a "rectangular" distribution, for reasons that are obvious in figure 3-6.

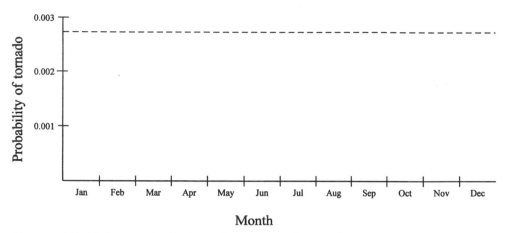

Figure 3-5. Uniform distribution for daily p(tornado) as described in problem 3-7a.

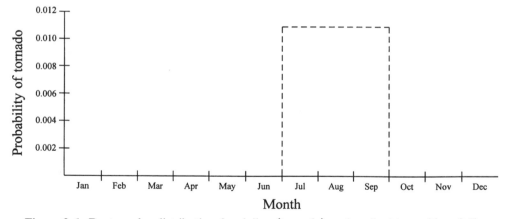

Figure 3-6. Rectangular distribution for daily p(tornado) as described in problem 3-7b.

Since each day in the summer months is equally likely, there is a 1/92 chance that a tornado will occur on July 4, and a 0 probability that it will occur on January 10.

Solution 3-7c

This case indicates a little more information. As in (*b*), the limits are known (July 1 through September 30), but now in addition it is known that August 15 is the most likely date. Consequently, a triangular distribution might be applicable, since there is no further information about the shape of the distribution. Figure 3-7 depicts this triangular distribution.

Finding the probability that a tornado will occur on July 23 requires two pieces of information. First, it is necessary to recall that the cumulative probability that the tornado will occur sometime between July 1 and September 30 is 1, so the cumulative probability up to August 15 (i.e., the probability that it will have happened by then) is 0.5. Second, the number of days up to August 15 is 46.

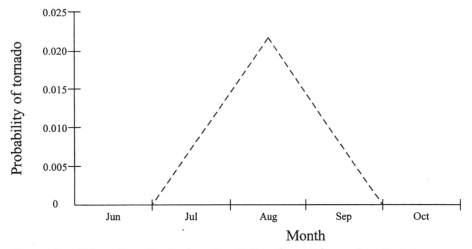

Figure 3-7. Triangular distribution for daily *p*(tornado) as described in problem 3-7c.

From these data, the probability that the tornado will have happened by August 23 can be calculated using the geometric identities that the lengths of the sides of a right triangle scale in direct proportion if the angles do not change, and that the area of a triangle is one-half of the product of the base and the height. The probability that the tornado will occur on August 15 is the height of the triangle, and 46 days is the base, so

$$\text{area} = 0.5 \times \text{base} \times \text{height},$$

$$= 0.5 \times 46 \times p(\text{Aug 15}).$$

Rearranging,

$$p(\text{Aug 15}) = 0.5/(0.5 \times 46),$$

$$p(\text{Aug 15}) = 0.022.$$

Since the base of the triangle for July 23 is $23/46 = 1/2$ that for August 15, the probability must be $1/2$ of that as well, so

$$p(\text{July 23}) = 0.5 \times p(\text{Aug 15}) = 0.5 \times 0.022$$

$$= 0.011.$$

The Binomial

The binomial distribution is frequently useful in risk assessment. The binomial applies when there must be one (and only one) of two states of nature for a given variable. For example, one can either have cancer or not have cancer, have a safe flight or an unsafe flight, and so on. A number of statistical tools have been developed specifically using the binomial, and here we will explore some of those.

Many risks can be modeled as binary systems, where either an event happens or it doesn't. Consequently, the standard example of a coin flip is a useful way to understand how the binomial works. Consider a coin with a 0 on one side and a 1 on the other. The probability of getting a 1 in a single flip is 0.5, and the probability of getting a 0 is likewise 0.5. So the expected value of a single flip is

$$E(\text{flip}) = 0.5 \times 1 + 0.5 \times 0 = 0.5.$$

This leads to a more general formula for the expected value of n flips:

$$E(n \text{ flips}) = n \cdot p,$$

and the standard deviation of a binomial variable is

$$s = n \cdot p(1 - p).$$

Given many trials, the shape of the binomial distribution converges on the normal. If n is large (and p is not very close to either zero or one), we can expect that in 90% of times we do n trials, the total value of the summed tosses will be within $\pm 1.96[n \cdot p(1 - p)]$.

The binomial distribution is also useful when we want to evaluate whether two different sets of trials are the same. For example, we might want to know whether a group of patients taking a medicine are less likely to get well than another group taking a placebo. Or, as in this case, we might want to decide whether the average radon level in one neighborhood is the same as that in the other.

Suppose that you are trying to decide whether the fraction of the houses with radon levels above 10 pCi/l (which you term "high" levels for this study) in one neighborhood is different from that in a second neighborhood. In the first neighborhood, there are 57 houses, 17 of which have high radon levels. In the second, there are 72 houses, 18 of which have high radon levels. Test the hypothesis that the two neighborhoods are the same in terms of the fraction of houses with high radon concentrations.

Solution

Here, we are testing the hypothesis that the fractions are the same, which is equivalent to saying that the mean of the differences between the two fractions is zero. The standard deviation of the differences between the two means is given by

$$S = \sqrt{q \times (1 - q)\left(\frac{1}{n(1)} + \frac{1}{n(2)}\right)}, \qquad \text{3-10}$$

where $q = [p(1)n(1) + p(2)n(2)]/[n(1) + n(2)]$, $p(i) =$ the probability that a sample from the ith group will take the value 1, and $n(i) =$ the number in the ith group. In this case, let "high" radon be 1, and "low" (more accurately, "not high") be 0; then

$$p(1) = 17/57 = 0.30,$$
$$n(1) = 57,$$
$$p(2) = 18/72 = 0.25,$$
$$n(2) = 72.$$

Then,

$$q = (17 + 18)/(57 + 72) = 0.27,$$

and

$$s(\text{differences}) = \sqrt{0.27 \times (1 - 0.27)\left(\frac{1}{57} + \frac{1}{72}\right)} = 0.079.$$

We can use this value to do a Z-test, using the equation

$$Z = \frac{p(1) - p(2)}{s(\text{differences})} = \frac{0.30 - 0.25}{0.079} = -0.63.$$

If we test the hypothesis that they are the same at a 95% confidence level, we see from appendix A that we would need a Z-score of -1.64, so we cannot reject the hypothesis that these two neighborhoods are the same.

Another feature of the binomial is that we can evaluate the uncertainty created by randomness associated with small sample size. We can use the known probability of occurrence and the sample size to estimate a binomial distribution around the sample size. Given the standard deviation and the expected value, we can modify equation 3-7b:

$$Z = \frac{\bar{x} - np}{\sqrt{np(1 - p)}} \qquad 3.7c$$

For example, in the first neighborhood above, the observed frequency is 0.30, and the standard deviation is 3.5. Consequently, if we wanted to know how many high-radon houses we could expect in another neighborhood of 90 houses with characteristics similar to those in neighborhood 1, we could expect $90 \times 0.30 = 27$, and a 95% confidence range would be $\{27 + 2s, 27 - 2s\} = \{34, 20\}$. Note that if we use this method to ask whether neighborhood 2 is similar to neighborhood 1, we find that an answer of 18 houses is within the 95% range.

Problem 3-8. Fitting a Model

This problem demonstrates the normal, lognormal, and exponential distributions. The normal has been proposed as a possible distribution for radon concentrations. How well do the sample data in table 3-6 fit the normal distribution?

Solution 3-8

Normal or Gaussian Distribution

Probably the most familiar distribution is the *normal* (alternatively, Gaussian or bell-shaped). Equation 3-11 shows the formula for this distribution, and figure 3-8 graphs the *standard* normal distribution, where $\mu = 0$ and $\sigma = 1$. The normal is useful in statistical analysis for the reasons described above.

$$y = \frac{1}{\sigma \cdot \sqrt{2\pi}} e^{-(x - \bar{x})^2 / 2\sigma^2}. \qquad 3\text{-}11$$

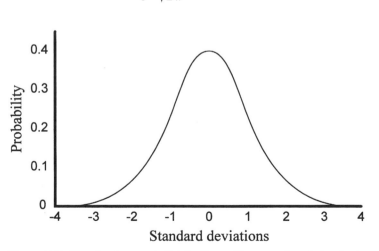

Figure 3-8. Standard normal distribution with mean 0 and standard deviation 1.

The main theoretical reason we might expect a distribution to take this form is if it is the sum of a number of independent events. Empirically, the normal is not often useful, since it is necessarily symmetric and is unbounded both above and below.

When we want to know whether a particular model (in this case, a distribution, but true as well for other models) represents the available data well, we test for goodness of fit. Such a test requires that we calculate expected values for the data, where the expectation is generated by the proposed model. We then compare some measure of the variation between data and expected values, typically using either a normalized difference or a ratio.

One tool that can be used to evaluate goodness of fit is R^2 (read "R squared"). R^2 values run from 0 (proposed model does not fit at all) to 1 (proposed model fits the data perfectly). It is calculated as follows:

1. Calculate the average value of the data,

$$\bar{y} = \frac{1}{n} \sum_{i=1}^{n} y_i.$$

2. Calculate and square the difference between each data point and the average value. Sum these differences. This is the total sum of squares (TSS).

$$\text{TSS} = \sum_{i} (y_i - \bar{y})^2.$$

3. Calculate and square the difference between each data point and the value you would expect it to take given the proposed model. Sum these differences. This is the error sum of squares (ESS).

$$\text{ESS} = \sum_{i} (y_i - y_i^e)^2,$$

where the superscript e refers to "expected."

4. Calculate R^2:

$$R^2 = 1 - \frac{\text{ESS}}{\text{TSS}}.$$

$0 \leq$ ESS $\leq$ TSS in all cases, because TSS is made up of two components, an error term (ESS) and a model or regression term (RSS), which constitutes the real difference among the data points. Intuitively, if the model fits every point exactly, ESS $= 0$, and $R^2 = 1$. If the model has no relation to the data, ESS $=$ TSS, and $R^2 = 0$. The R^2 is sensitive to absolute values of numbers and is most appropriate for fitting linear models. Figures 3-9a and 3-9b depict the values used to calculate R^2 for linear and nonlinear functions.

Chi-Squared Distribution

Another tool often used to calculate the fit of a model to a data set is the chi-squared (χ^2) test. This test is similar in philosophy to the t-test, since it compares how data points vary around an expected value. However, while the t-test is limited to comparing a single calculated value to a single hypothesized value, the χ^2 allows the calculated value to vary along the hypothesized distribution, so that the entire equation can be compared to a known data set. In addition, the χ^2 test corrects for the absolute value of each data point, and so is not sensitive to large and small values.

The general formula for the χ^2 is

$$\chi^2 = \sum_{i=1}^{n} \frac{(x_i^o - x_i^e)^2}{x_i^e}. \tag{3-12}$$

The superscript o refers to observed value, and the e refers to expected (hypothesized) value. The intuition behind this is that we are looking at the squared (and thus non-negative) difference between what the data actually are and what we would expect the data to be if they were taken from the distribution. Dividing by x^e "normalizes" to the size of the data, so that we may use one χ^2 table whether the numerical values are generally large or small.

For many distributional models, we have equations that describe the probability of particular values coming up; for the normal we use equation 3-11. The expected value for a particular data point is the number of data points times the probability of that data point coming

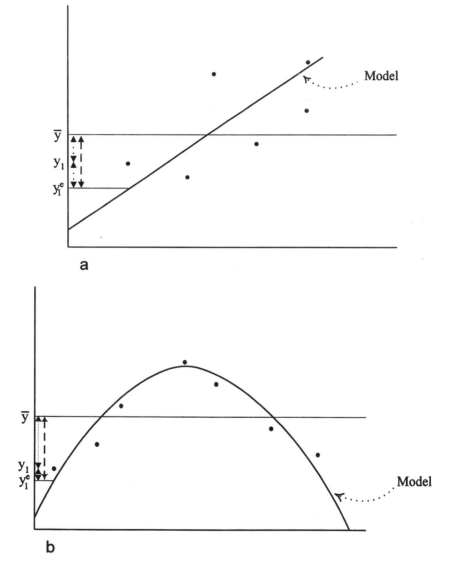

Figure 3-9. (a) $\bar{y}$, y_i^e, and y_i for a linear regression. (b) $\bar{y}$, y_i^e, and y_i for a nonlinear model.

up in a single random draw, or $n \cdot p_i$. Therefore, the formula for calculating the test statistic for a distribution is

$$\chi^2 = \sum_{i=1}^{n} \frac{(x_i^o - n \cdot p_i)^2}{n \cdot p_i}.$$

3-13

To calculate this for the large data set, use equation 3-11 to calculate the probability that each observed value will come up, given the normal distribution. Next, sort the data into "bins" of equal value and count the number of points in each bin. In this case, since there are 43 data points ranging from 0.74 to 6.48, six bins ranging from 0.6 to 6.6 is reasonable: too many bins and the distribution will appear to be uniform; too few bins and resolution will be lost. Table 3-8 shows the bins, while figure 3-10 shows the bins graphically, with the normal distribution superimposed.

Now we can do a χ^2 test for the actual number of houses at each concentration level versus the number expected if they are distributed normally:

$$\chi^2 = \frac{(3 - 1.8)^2}{3} + \frac{(3 - 5.6)^2}{3} + \frac{(12 - 10.6)^2}{10.6} + \frac{(11 - 12.1)^2}{11}$$

$$+ \frac{(9 - 8.2)^2}{9} + \frac{(5 - 3.4)^2}{5}$$

$$= 3.48.$$

Table 3-8 Number, Probability, and Expected Number of Houses Sorted by Bin

Bin	0.6 to 1.6	1.6 to 2.6	2.6 to 3.6	3.6 to 4.6	4.6 to 5.6	5.6 to 6.6
Actual number of houses	3	3	12	11	9	5
Expected probability	0.041	0.13	0.25	0.28	0.19	0.078
Expected number of houses	1.8	5.6	10.6	12.1	8.2	3.4

Note: Expected probability is based on the probability of the midpoint of each bin, based on a normal distribution, mean = 3.85, $s = 1.4$.

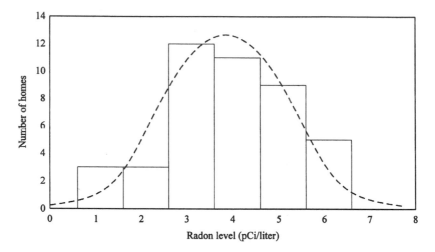

Figure 3-10. Normal distribution with mean 3.85 and standard deviation 1.4, superimposed on the number of houses in each bin, as described in problem 3-8.

Once you have calculated the χ^2 test value, you can compare it to the values in appendix C. Recall that the chi-squared value will increase as the difference between the hypothesis and the data increases, so a high chi-squared value tells us to reject the hypothesis that is being tested. Since the value from the table of chi-squared (6 degrees of freedom, 0.90) is $\chi^2 = 10.6$, and the calculated chi-squared value for this data set is $\chi^2 = 3.48$, we cannot reject the hypothesis that the data are normally distributed at a 90% confidence level.

Refer to the table of chi-squared values in appendix C. There are two salient points. First, as the number of degrees of freedom goes up, so do the chi-squared values necessary to find a significant difference. Recalling the importance of sample size, this should not be surprising, since it is very hard to say something meaningful from a very small data set, or when freedom is highly restricted. Second, the greater the desired certainty level, the greater the necessary chi-squared value. Again, since this is a sum of squared observed differences, the greater the sum, the greater the chance of there being a real difference.

Because the distribution of risks varies significantly among different distributional forms, the policies suggested by other distributions are often very different than those suggested by the normal.

Problem 3-B. R^2 versus χ^2

Calculate the R^2 for the sample data set. Compare this to the χ^2.

Problem 3-C. How Many Bins?

Suggest (and justify) an alternate set of bins for the data in table 3-6 and recalculate the χ^2. How sensitive is the χ^2 to selection of bins?

Problem 3-9. Distributional Models

    a. Radon concentrations in U.S. houses are not normally distributed. Figure 3-11 shows an estimate of the actual distribution, based on data for > 1000 households in 38 different locations in the United

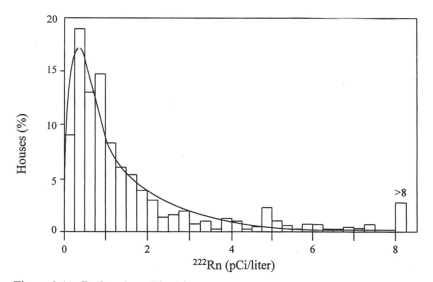

Figure 3-11. Radon data. The bins represent the percentage of houses at a given range of radon levels. The overlay is the lognormal distribution with GM = 2.84 and GSD = 0.96 (pCi/liter). From Nero et al. 1986.

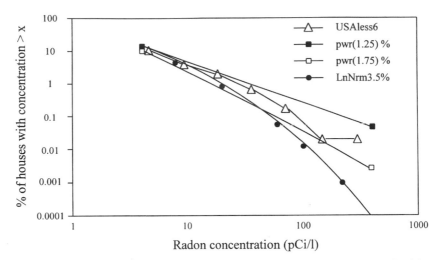

Figure 3-12. Comparison between Terradex data set "USA less 6 states" with theoretical distributions for high radon levels. After Goble and Socolow (1990), figure 3B. The Terradex data set contains information on household radon concentrations for forty-four of the states.

States. What are some theoretical reasons why the lognormal distribution might be appropriate?

b. Most experts on radon risk believe that household radon concentrations in the United States appear to be lognormally distributed, with a mean of 1.5 pCi/l and a standard deviation of 2.6 pCi/l (see, for example, Nero et al. 1986). In 1990, Goble and Socolow suggested that a power law distribution, with exponent 1.25, fits the data just as well. Figure 3-12 shows the data points, along with the lognormal and exponential distributions, as presented by Goble and Socolow. Why might you select one distribution over the other, and what is the significance of the difference?

Solution 3-9

Solution 3-9a. Lognormal Distribution

While the normal distribution appears to fit the sample data set adequately, data on household radon in most studies do not fit the normal distribution well. This is expected for both theoretical and empirical reasons.

The theoretical reasons are as follows: First, radon levels cannot be negative; consequently, even if the data were to follow a generally normal shape, it would have to be truncated. Second, the generation of radon gas is a complex phenomenon and does not appear to be the sum of independent random variables. The presence of radon in a house is not an independent event; rather, it is generated by a number of interrelated events, such as the decay of uranium into radon and other radioactive isotopes (see figure 2-2), geography, building structure, and other factors.

Empirically, figure 3-11 shows the data on household radon concentrations for thousands of homes in the U.S. (Nero et al. 1986). It clearly is not symmetric; rather, the data are clustered around a central area and have a wide tail as radon levels get very high.

The lognormal distribution may provide a better fit to some variables than does the normal. This distribution arises when the logarithm of the variable is normally distributed. Since we are working with logarithms, the values will always be greater than zero. The main theoretical reason to expect that a variable might take such a distribution is that it is the product of several random variables. The lognormal often provides a good description of a distribution drawn from a physical set of objects, such as the number of vehicles traveling at different speeds or particle sizes. This is true because the lognormal is both positive and has a long, low-probability tail, which describes the rare elements of a set, such as one driver doing 120 miles per hour when most are driving at 65.

The lognormal distribution is empirically useful because it cannot take nonpositive values and may simply fit the data well. Equation 3-14 is the formula used to calculate the lognormal distribution.

$$y(x) = \frac{1}{x\phi\sqrt{2\pi}} e^{-(\ln(x)-\bar{x})^2/2\phi^2},$$ 3-14

where ϕ is the mean,

$$\phi = \left(\frac{1}{n} \sum_{i=1}^{n} \left(\ln x_i - \overline{\ln(x)} \right)^2 \right)^{1/2}.$$

The shape of the lognormal distribution is determined entirely by the mean and the standard deviation, which is what concerns Goble and

Socolow. Since the shape is predetermined, the width of the tails is prescribed, which Goble and Socolow (1990) argue might significantly underestimate the number of homes with very high radon levels.

Solution 3-9b. Exponential Distribution

The exponential or power rule function is often used to describe the amount of time between two events. As such, it is especially useful for evaluating how likely two independent failures are to occur within a given period of time. If either is tolerable if it happens alone but both together are not tolerable, a model of the likelihood of the intolerable coincidence is useful. Like the lognormal distribution, the exponential can be useful simply if it fits the data well. Goble and Socolow find that an exponent of 1.25 fits the available data as well as does the lognormal used by Nero and others, which suggests quite different policy measures. Figure 3-12 is taken from Goble and Socolow: it depicts the data points, along with both lognormal and exponential curves. The exponential's wider tail suggests that there may be a much larger number of high-level homes than predicted by the lognormal. If this is true, society would be prudent to put a relatively larger fraction of resources into identifying and mitigating these high-risk homes.

Problem 3-D. Fitting the Lognormal Distribution

Calculate the geometric mean and geometric standard deviation for the sample data in table 3-6. Fit the data to a lognormal model, evaluate the goodness of fit for this model, and compare your results to the fit of the normal distribution model.

Problem 3-E. Dealing with Grouped Data

From figure 3-11, estimate the value of each bin and fit it to a lognormal distribution with mean 0.96 and standard deviation 2.84. Do this first excluding the final bin, which represents all of the data in the tail, from 8 to ∞, and then find a method to include the final bin.

References

Brill, A. B., Becker, D. V., Donahoe, K., Goldsmith, S. J., Greenspan, B., Kase, K., Royal, H., Silberstein, E. B., and Webster, E. W. (1994). "Radon update: facts concerning environmental radon levels, mitigation strategies, dosimetry, effects and guidelines." SNM Committee on Radiobiological Effects of Ionizing Radiation. *Journal of Nuclear Medicine* 35(2):368–85.

Goble, R. and Socolow, R. (1990). "High radon houses: Implications for epidemiology and risk assessment." *Cented Research Report No. 5.* Worcester, MA: Clark University.

Larsen, R. J. and Marks, M. L. (1986). *An Introduction to Mathematical Statistics and its Applications.* 2nd Ed. Englewood Cliffs, NJ: Prentice Hall.

NAS-BEIR IV. (1988). Health Risks of Radon and other Internationally Deposited Alpha-Emitters. *Committee on the Biological Effects of Ionizing Radiations.* National Research Council, Washington, DC: National Academy of Sciences Press.

Nero, A. V., Schwehr, M. B., Nazaroff, W. W., and Revzan, K. L. (1986). "Distribution of airborne radon-222 concentrations in U.S. homes." *Science* 234:992–97.

Nero, A. V., Gadil, A. J., Nazaroff, W. W., and Revzan, K. L. (1990). *Indoor Radon and Decay Products: Concentrations, Causes, and Control Strategies.* DOE/ER-0480P. Washington, DC: U.S. Department of Energy.

U.S. EPA (1992). Technical Support Document for the 1992 Citizen's Guide to Radon. EPA 400-R-92-011 (May).

Wonnacott, Thomas H. and Wonnacott, Ronald J. *Introductory Statistics for Business and Economics.* 4th Ed. p. 518. New York: John Wiley and Sons.

4

Uncertainty, Monte Carlo Methods, and Bayesian Analysis

Introduction

This whole book is really about uncertainty. In this chapter we focus on several methods to evaluate and characterize uncertainty and to judge its impact on decisions.

If we knew the answer exactly to the question, "Will this particular airplane crash?" there would be no need for a risk analysis, there would only be a need to act to correct or avoid the problem. If we knew exactly the probability of the airplane crashing but could not fix it, then the only question would be, "Are we comfortable with this level of risk?" In this case our decision is simply a question of first, statistics, and second, policy. The real world is obviously not so simple, nor so uninteresting.

There are several types of uncertainty, and an important task in risk analysis is to determine what kinds of uncertainty are likely to affect your findings. With random error, (alternatively called statistical uncertainty), a system has one or more unpredictable components, but based on your understanding, the system can be characterized. Rolling an honest die is an example—you don't know what the outcome of a roll will be, but you can say that there is a one in six chance that a given roll will be a 4, and that the most likely sum of two roles is 7. Chapter 3 on statistics provides the basics of this type of analysis, which is also repeated in varying degrees of complexity throughout the rest of this book.

The second type of uncertainty is more insidious, and is the focus of this chapter. What if there is a consistent bias in the system or model? How can such biases be identified, characterized, and (hopefully) understood? If you notice, for example, that the die seems to be coming up 6 particularly frequently, what is happening? As we found in chapter 3, first, you may need more data to determine if the frequency of 6s is inconsistent with chance. It is possible that the die is weighted, in which case enough observations may lead you to an understanding of the bias, or enough observations of the die itself may reveal that 6 is painted on more than one face, or that a bump on one side causes the die to land one way more than another. Here you have discovered a structural explanation of the bias, which can then become part of your predictive model.

Alternatively, what if the problem is not the die at all, but the observer, or the method of observation, or the equipment? If you watch the die through dirty glasses and cannot distinguish between 3 and 5, then there is a systematic uncertainty that not only alters the results as you record them, but also may be particularly difficult to discover. Systematic error is defined as the difference between the true value of the quantity you are observing and the value that you report through observations.

To test the die, however, you will want both to try out a large number of rolls and also to roll it in different ways, for example, by throwing it as in craps or slapping a cup of dice down as in backgammon. The question is then, does throwing the die in different ways matter, and how can you explore the results? This may sound like just spending a lot of time in a casino, and in fact the computational method that mimics this is called the Monte Carlo method.

The analysis of systematic uncertainty puts the observer in as much question as the system under study: the observer is part of the system. While this makes sense in the abstract, in practice the observer is often the most difficult part of the system not only to characterize but to think about clearly and objectively.

Repeated throughout the problems in this book is the admonishment to "be explicit about your assumptions and the areas of uncertainty." As should be clear by now, risk analysis is rife with uncertainties: Is the right model being used? Is the model designed correctly? How good are the data? How valid are the assumptions? To this list we need to add methods for characterizing the observer. Uncertainty must be acknowledged, identified, or evaluated even when quantifying statistical or systematic uncertainty is difficult or impossible. This chapter examines some techniques for identifying biases, and methods to recognize and characterize them. First we will look at some cases where uncertainty can dominate the analysis to illustrate a variety of ways in which systematic uncertainty can be addressed. Subsequent problems introduce Bayesian analysis and Monte Carlo techniques.

Problem 4-1. Measuring the Speed of Light

One particularly intractable aspect of systematic errors is that they can be unexpectedly subjective. Many policy decisions are clearly subjective, or value based—such as whether to fly in an airplane with unknown accident risk. In contrast, this problem considers a case where we would expect the approach to be objective: historical measurements of a basic physical constant, the speed of light.

Plotted in figure 4-1 are the historical measurements of the speed of light, along with the total errors (at the one standard deviation level) reported by the experimenters. For reference, the dashed horizontal line is the currently accepted value of the speed of light. The gray line is the value recommended by the U.S. Office of Measurements and Standards.

    a. Provide a rough quantitative measure for the degree of systematic error in these measurements.

    b. Develop a hypothesis for the clustering of the pre-1940 measurements at a value inconsistent with the currently accepted value.

Solution 4-1

Solution 4-1a

The error bars for each measurement are for one standard deviation; thus just under 70% of the time, you would expect the error bars to intersect the "true" value of the speed of light.[1] Reading off the graph, of the 26 measurements, the error bars for 13 intersect the true value. This is a small sample, but it is clear that something is amiss.

---

[1] "True" in this context means "currently accepted." It is altogether reasonable to expect that, in fifty or a hundred years, the values we now consider "true" for some parameter will be considered as far off the mark as we consider those of the past!

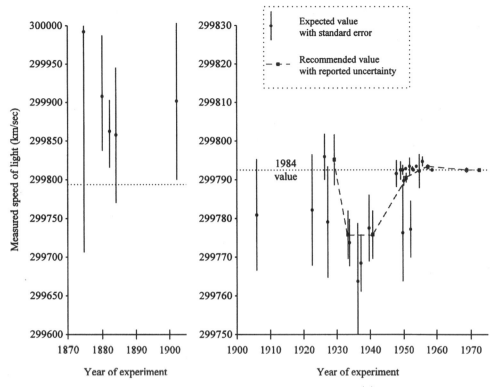

Figure 4-1. Historical series of measurements of the speed of light (*c*) with the accompanying experimental error. The accepted value of *c* (the 1984 value of $2.99792456 \pm 0.000000011 \times 10^8$ m/sec) is indicated as a dashed line. After Henrion and Fischhoff 1986.

Solution 4-1b

The error bars for the cluster of measurements made between 1930 and 1940 all fail to intersect the true value of the speed of light. The fact that this cluster of measurements are all "low" while the set of earliest measurements (in the 1800s) are all "high" suggests that more is going on here than statistical fluctuation.

The simplest explanation of the clustering of measurements in the 1930s concerns the equipment. It is likely that many of these experiments used a similar apparatus, possibly refinements of a particular method, and certainly the electronics and other components all relied on technology of the same vintage. In such a case, bias in one experiment is likely to be echoed in other experiments that are similarly constrained by technology. The dashed line in figure 4-1 showing the recommended value of the speed of light shows the trend in the accepted values most clearly. Other fundamental constants (the mass of the electron, the fine structure constant, Planck's constant, and Avogadro's number) all show the same trend (Henrion and Fischhoff 1986).

There has been a great deal of research on the estimation and calculation of systematic errors (for summaries, see Morgan and Henrion 1990 and Shlyakhter et al. 1994). A universal finding is that experts and nonexperts alike tend to underestimate errors, that is, they are overconfident in their results. One reason is clear: small effects that are poorly understood are likely to be minimized or ignored, and some important errors may not be known at all to researchers of a given era.

There are sociological reasons why estimates may cluster as well. Science is the business of skeptical investigation: new effects are challenged and must meet higher standards of evidence before they are accepted than would a repeat of a known result with conventional methods and equipment. This makes investigative research conservative in accepting new findings. A healthy dose of skepticism is a useful filter on new and unverified or unreplicated findings. Once a finding is accepted, however, the systematic errors that accompany the measurement become endemic.

Now that we are alert for biases in measurements, it is interesting to extend this skepticism to forecasts and other assessments of probabilities and risks. Forecasting is a frequent goal of risk analysis. If systematic uncertainty appears in the measurement of fundamental physical constants, then it is likely to be rampant in highly subjective efforts to forecast future trends in a variety of parameters. We can explore this with any set of forecasts for which we now have the true result, such as stock market returns (Gordon and Kammen 1996), population, or other demographic trends (Shlyakhter et al. 1994).

Problem 4-A. Energy Forecasts

Consider the set of forecasts in figure 4-2 for total U.S. energy use in the year 2000. The estimates connected by bars are high and low

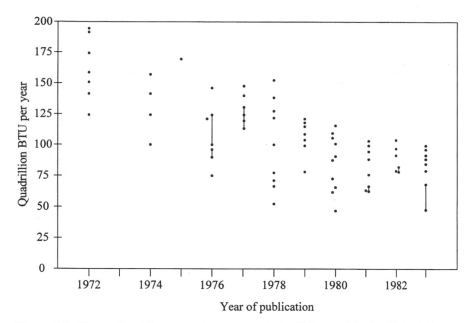

Figure 4-2. Forecasts of the amount of energy that will be used in the United States (in quads, or quadrillion BTUs) in the year 2000 plotted against the year in which the predictions were made. The cases where two forecasts are connected by a line indicate "high" and "low" estimates made by the same group. The forecasts come from an ideologically and politically broad range of organizations, including the Department of Energy, the American Petroleum Institute, Mobil Oil, and the Rocky Mountain Institute. After Goldemberg et al. 1987.

estimates by one organization (such as Shell Oil or the U.S. Department of Energy). Most forecasts are point estimates, but you can make some reasonable assumptions to estimate a confidence interval around the estimates provided. For example, you could assume that each estimate is the median from a normal distribution, and calculate the distribution for estimates singly, or in groups averaged over an interval of a few years. Look up the current total energy consumption (1998, in this case), and use that for the year 2000.

   a. Develop a probability distribution to measure the discrepancy between the forecast values and the true energy use. One convenient measure is to look at the probability that the actual value fell outside the confidence interval associated with each estimate. You expect some fraction of the estimates to be inconsistent with the actual value. How many, and by how much?

b. If the forecasts were all independent, you would expect the shape of the curve found in (a) to be a normal distribution. Is it? If not, are the forecasters overconfident as we have hypothesized?

Problem 4-B. Forecasting the Impacts of Climate Change

Economist William Nordhaus (1994) conducted a survey of twenty researchers in the area of global change science, economics, politics, ecology, and engineering. Each was asked to estimate the impacts of a three-degree global warming by 2090 on the global product. A comparison of the median and 10% and 90% confidence intervals for each scholar is listed in figure 4-3. The estimates are listed by the code number for each respondent. The estimates by natural scientists were generally 20–30 times higher than those made by the economists: natural scientists estimated the chance of large consequences to be over 10%, while economists on average assigned a 0.4% chance.

a. What does this set of estimates say about current understanding of the potential impacts of global change?

b. Construct an estimate of potential costs of global warming from the forecasts by these experts. What are the benefits and problems of such an estimate?

Bayesian Statistics

Both stochastic and structural uncertainty in a probabilistic risk model or a decision process necessitate that the state of information be characterized. A useful framework for such characterization is one where expected outcome probabilities are explicitly recognized. The Bayesian view of probability fits well with this requirement.

Bayesian analysis recognizes the attention that systematic bias directs to the observer. Instead of thinking of probability as simply the frequency of an event, in the Bayesian or "personalist" view, a probability is a construction that depends on the information or "state" of the observer. As new information is gathered, the observer may revise the

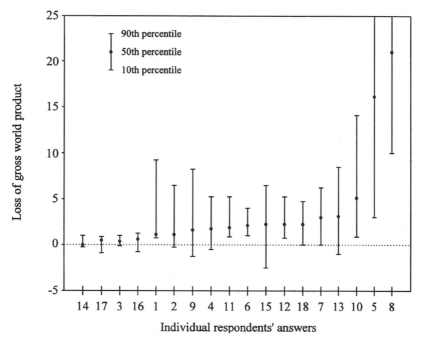

Figure 4-3. A set of estimates by prominent economists, ecologists, and natural scientists estimating the loss of gross world product that is likely to result from a doubling in the atmospheric concentration of carbon dioxide by the mid twenty-first century. This release is anticipated to increase global mean temperatures by 3°C by 2090.

understanding of the system, thereby explaining some statistical or systematic biases and possibly revealing others.

For example, if the observer begins by assuming that a die is honest, then equal probabilities for each number are expected and it would come as no surprise if a series of rolls revealed this pattern. Not expecting a weighted die, a high frequency of 6s is likely to be a surprise. If, however, the observer *knows* that the die is weighted—perhaps having tampered with it herself—she will not be surprised at all to see 6 on many of the rolls. The Bayesian observer would continue to adjust her expectations based on the new information that came in: first, the string of 6 rolls, and then, possibly, information that the die was weighted. She would adjust the probability distribution (the chance of getting a $1, 2, \ldots, 6$) to reflect this additional information.

Bayes' theorem provides a framework to evaluate probabilities, or risks, based on different sets of information. We will use the standard notation that the probability of an event X given information H is $P(X \mid H)$, so that

P(I will get a tropical disease in the next year | I am a 36-year-old male)

could be used for one estimate, but we could presumably make a better estimate if told that I am a 36-year-old male who

- Works in the tropics on environment and health issues and is exposed to sick people
- Takes the appropriate medicines
- Has a healthy immune system, and so on.

Now we have a more complete set of information, $P(X \mid H_i)$, that determines the probability of X given a set of facts, H_i.

Formally, the probability of X given a set of observations can be written as an intersection of sets of facts, so that the *conditional probability* is defined as

$$P(A \mid B) = \frac{P(A \cap B)}{P(B)}, P(B) \neq 0. \qquad \text{4-1}$$

The intersection of A and B ($A \cap B$) is an important concept in terms of probabilities and leads to a number of obvious statements:

1. In the definition of $P(A \mid B)$ in equation 4-1, we have a simple statement of the fact that a probability must have a value between 0 and 1:

$$0 \leq \frac{P(A \cap B)}{P(B)} \leq 1.$$

2. If B is a subset of A, then $A \cap B = B$, and then $P(A \mid B) = 1$.
3. If A and B are disjoint, with no overlap, then

$$A \cap B = \varnothing,$$

which implies that $P(A \mid B) = 0$.

Now we can think more generally about how knowledge of the probability of one event impacts our knowledge of another event and therefore arrive at Bayes' theorem.[2]

The probability of an event B, $P(B)$, is determined by the event B in relation to a set of facts, A, and their associated probabilities, so that the probability of B is

$$P(B) = \sum_{i=1}^{n} P(B \cap A_i).$$

In forming these conditional terms, $B \cap A_i$, we assume that each of the events or facts, A_i, is unique, namely, that $A_i \cap A_j = \varnothing$ if $i \neq j$.

$P(B)$ can be rewritten using the definition of conditional probability above, in a form that is called the *total probability theorem*:

$$P(B) = \sum_{i=1}^{n} P(B \mid A_i)P(A_i). \qquad 4\text{-}2$$

We can now return to the statement of conditional probability, and rewrite it with the terms interchanged:

$$P(B \mid A) - \frac{P(B \cap A)}{P(A)}, \, P(A) \neq 0. \qquad 4\text{-}3$$

If we now substitute for $P(B \cap A)$ from the total probability theorem we have a form of Bayes' theorem:

$$P(B \mid A)P(A) = P(A \mid B)P(B).$$

But now if we look at any event, A_i, with respect to B, we have

$$P(A_i \mid B) = P(B \mid A_i)P(A_i)/P(B).$$

Perhaps the most useful operational form of Bayes' theorem can then be written by substituting into this equation the total probability theo-

---

[2] This derivation follows that of Williams (1991).

rem (equation 4-2), which yields

$$P(A_i \mid B) = \frac{P(B \mid A_i)P(A_i)}{\sum\limits_{i=1}^{n} P(B \mid A_i)P(A_i)}.$$

4-4

We will use this form for calculations in the next few problems (and later in the book) because each term can be easily determined.

What can be done with this? Consider first the following problem.

_____

Problem 4-2. Interpreting Test Results

Suppose that you are being screened for a disease that afflicts one person in five thousand, and that there are no symptoms until it is too late for a cure. The screening test has a 5% false positive rate and a 2% false negative rate.

a. What do you learn when told that you are fine and do not have the illness?
b. What do you learn when told that you should see your doctor?

Solution 4-2

Solution 4-2a

Based on the simple probability, the chance of being well or ill, *in the absence of any additional information*, is

$$P(\text{no disease}) = 1/5,000 = 0.9998,$$

$$P(\text{diseased}) = 0.0002.$$

But now using the information on the test probabilities, we have additional information, or priors:

$$P(\text{positive test} \mid \text{no disease}) = 0.05,$$

$$P(\text{negative test} \mid \text{no disease}) = 0.95,$$

$$P(\text{positive test} \mid \text{diseased}) = 0.98,$$

$$P(\text{negative test} \mid \text{diseased}) = 0.02.$$

Each of these $P(\dots \mid \dots)$ values is a probability associated with the type I and type II errors we discussed in chapter 3. The *false positive*, where the test indicates a disease when none is present (here a 5% rate) is a type I error. A type II error, or a *false negative*, occurs when the disease is present, but the test indicates a healthy individual (here a 2% rate). To compare alternatives we will adopt from Schmitt (1969) a convenient tabular framework using Bayes' theorem.

The rows in tables 4-1 and 4-2 should 'read' to you like a multiplication of the individual terms in Bayes' theorem. The first four columns provide the numerator:

$$\text{prior} \times \text{probability } P = \text{joint probability.}$$

The right arrow $\rightarrow$ indicates that this total should then be normalized (divided) by the product of probabilities from the denominator of Bayes' theorem, namely,

$$\sum_{i=1}^{n} P(B \mid A_i)P(A_i).$$

Table 4-1 Tabular Method for Applying Bayes' Theorem: Positive Test Result. A tabular approach can be a convenient way to organize the calculation of joint and posterior probabilities using Bayes' theorem.

Alternative	Prior		P (positive | alt)		Joint		Posterior
Not diseased	0.9998	$\times$	0.05	$=$	0.04999	$\rightarrow$	0.9961
Diseased	0.002	$\times$	0.98	$=$	0.00196	$\rightarrow$	0.0039
Sum	1.0000				0.050186		1.0000

Table 4-2 As in Table 4-1, but for Negative Test Result

Alternative	Prior		P(positive | alt)		Joint		Posterior
Not diseased	0.9998	$\times$	0.95	$=$	0.949810	$\rightarrow$	0.999996
Diseased	0.002	$\times$	0.02	$=$	0.000004	$\rightarrow$	0.000004
Sum	1.0000				0.949814		1.0000

Thus, reading across the rows in the table computes the Bayesian probability:

$$\text{prior} \times \text{probability } P = \text{joint probability} \rightarrow \text{posterior.}$$

Table 4-1 summarizes what we know from Bayes' theorem if the test result is positive, and table 4-2 does the same for a negative test result. From these tables, we see that even with a positive test result, the chance of actually having the disease is 0.0391, or one in 256, while given a negative, the chance that you have the disease is not just one in 5,000 but one in 250,000.

Bayesian analysis can be, and is, applied widely in technical, medical, financial, and other types of probabilistic analysis. To get a flavor for the diversity of applications, consider a problem of labor unrest and potential financial losses.

_____

Problem 4-C. Bayesian Experts

A company must purchase a large quantity of a part either today or tomorrow. The price today is $14.50. The firm thinks that the price tomorrow will be either $10.00 or $20.00, and the chance of each price is the same.

A commodity market "expert" offers to provide the firm his forecast of tomorrow's price of either $10 or $20. The firm researches the expert's track record, and finds that he is correct 60% of the time. The expert's fee is $0.15 per unit of the commodity purchased. Should the company hire the expert?

_____

Problem 4-3. Bayesian Analysis of Radon Concentrations

The method and data discussed here are taken from Price et al. 1996.

The concentration of radon in U.S. homes varies across the nation. Although any house in the country could have a high level of radon, it would be helpful to identify regions in the U.S. in which the risk of a high household radon concentration is the greatest, so that the residents of these areas could be informed. Unfortunately, there are only sparse data available for many parts of the country. In this problem, we apply Bayesian analysis to calculate the best estimate of radon concentrations in the state of Minnesota based on the limited data available.

The information comes from a statewide survey performed during 1987–88 by the Minnesota Department of Health. Homes were tested throughout the state, and the concentrations followed a lognormal distribution at both the state and county levels. The geometric means and the number of measurements made in each county can be found in table 4-3. The geometric standard deviations for the counties were all approximately 2.1 bq/m^3. (Refer back to figure 2-3 for a map of all U.S. counties.)

a. Suppose that you want to estimate the true geometric mean for each county using the data in table 4-3. One way to accomplish this would be simply to consider the observed geometric means as the estimates of the true means. Explain why the small number of measurements for several counties might lead to an overestimate of the range of geometric means in the state.

b. Use Bayes' theorem to derive an estimate for the true value of the geometric mean (GM) of a county with the following assumptions:

   • The true county GMs are lognormally distributed (i.e., the values of log(GM) are drawn from a normal distribution with unknown mean μ and unknown variance σ^2).

   • The observations within a county are also lognormally distributed (i.e., the logarithms of the observations are drawn from a normal distribution with a mean equal to the true value of log(GM) and unknown variance κ^2). Let κ^2 be constant for all counties.

c. Estimate the necessary parameters from the data in table 4-3 and calculate point estimates for the true GM based on the expression derived in the previous section.

Solution 4-3

Solution 4-3a

Using the observed geometric means for estimates of the true values would be fine if each data point had a large number of observations behind it. However, many of the county GMs were based on one to five

Table 4-3 Measured Radon Concentrations in Each of the Counties in Minnesota and the Number of Measurements Made. The measurements were made in the basements of homes selected from each county. Note that in some counties quite a few measurements were taken, and the observed geometric mean (GM) is therefore likely to be reasonably representative. In the majority of the countries, however, there are few measurements (less than 10), and therefore a few very high or very low radon concentrations can dramatically bias the county GM.

Name	Number of observations	Observed GM ($Bq\ m^{-3}$)
Aitkin	4	73
Anoka	52	88
Becker	3	107
Beltrami	7	121
Benton	4	130
Big Stone	3	169
Blue Earth	14	250
Brown	4	189
Carlton	10	96
Carver	6	144
Cass	5	151
Chippewa	4	210
Chisago	6	107
Clay	14	222
Clearwater	4	100
Cook	2	73
Cottonwood	4	97
Crow Wing	12	97
Dakota	63	137
Dodge	3	224
Douglas	9	194
Faribault	6	75
Fillmore	2	105
Freeborn	9	259
Goodhue	14	235
Grant	0	NA
Hennepin	105	136
Houston	6	172
Hubbard	5	85
Isanti	3	107
Itasca	11	95
Jackson	5	280

Table 4-3 (*Continued*)

Name	Number of observations	Observed GM ($Bq\ m^{-3}$)
Kanabec	4	128
Kandiyohi	4	291
Kittson	3	115
Koochiching	7	57
Lacquiparle	2	498
Lake	9	54
Lake Of The Woods	4	168
Le Sueur	5	185
Lincoln	4	314
Lyon	8	242
McLeod	13	118
Mahnomen	1	145
Marshall	9	127
Martin	7	100
Meeker	5	126
Mille Lacs	2	69
Morrison	9	109
Mower	13	183
Murray	1	448
Nicollet	4	323
Nobles	3	255
Norman	3	103
Olmsted	23	126
Otter Tail	8	144
Pennington	3	78
Pine	6	72
Pipestone	4	200
Polk	4	146
Pope	2	134
Ramsey	32	112
Red Lake	0	NA
Redwood	5	234
Renville	3	156
Rice	11	221
Rock	2	137
Roseau	14	131
St. Louis	116	83
Scott	13	181
Sherburne	8	111

Table 4-3 (*Continued*)

Name	Number of observations	Observed GM ($Bq\ m^{-3}$)
Sibley	4	129
Stearns	25	148
Steele	10	181
Stevens	2	222
Swift	4	100
Todd	3	164
Traverse	4	231
Wabasha	7	208
Wadena	5	103
Waseca	4	62
Washington	46	131
Watonwan	3	344
Wilkin	1	344
Winona	13	163
Wright	13	182
Yellow Medicine	2	122

Source: Price et al. 1996.

measurements, far too few to accurately represent the true means. Some of the observed means will be lower than the true means, and some of the observed means will be higher. When they are combined to form the distribution of all GMs, it will be wider than the true GM distribution. This is always the case when dealing with statistics; any number of observations only approximates the hypothetical true value of the quantity being measured. With so few measurements, though, this effect can become a significant barrier to interpreting the data.

Solution 4-3b

We have the following data: n different observations of the logarithm of the radon concentration, $y_1, y_2, \ldots, y_n$, which are drawn from a normal distribution with a variance σ^2. From this we need to find the posterior estimate and distribution for the true county mean, log(GM). Using the notation that $N(\mu, \sigma^2)$ is a normal distribution, and θ is the true value of the logarithm of the geometric mean of the radon level in a particular county, we can write Bayes' theorem as

$$P(\theta, y) = P(\theta) \times P(y \mid \theta).$$

But we know that

$$P(\theta) = N(\mu, \kappa^2)$$
$$= k \times e^{-1/2\sigma^2(\theta - \mu)^2}.$$

$P(y \mid \theta)$ is a product of exponential terms, one for each county. So for a particular county,

$$P(y_i \mid \theta_i) = k_i \times e^{-1/2\kappa^2(y_i - \theta_i)^2}.$$

To solve, first multiply the $P(y_i \mid \theta_i)$s together and take the product with $P(\theta)$. It is now convenient to use the following definitions:

$$\mu = \overline{GM} = \text{mean of all observed GM values,}$$
$$e^{\sqrt{\kappa^2}} \approx GSD = 2.1,$$
$$\sigma^2 \approx 0.11.$$

Here GSD is the geometric standard deviation. After some algebra substituting back $y = \log(GM^{obs})$, as $P(y_n)$ can be rewritten

$$\log(GM^{est}) = \frac{(1/\sigma^2)\mu + (n_i/\kappa^2)\log(GM^{obs})}{(1/\sigma^2) + (n_i/\kappa^2)}, \qquad \text{4-5}$$

where GM^{est} is the estimated value of the GM, n_i is the number of measurements made in the county, and GM^{obs} is the observed value for the GM.

Entering these values into equation 4-5 yields estimates for the true GMs in each county, provided in table 4-4; the actual and estimated GMs are compared in figure 4-4. Note that the distribution has become narrower: observed GMs below the mean GM have been adjusted upward, and observed GMs above the mean have been moved downward.

Bayes' theorem can be used in a variety of situations where the information or process that informs the assignment of a probability is in

Table 4-4 Best-Estimate GMs for Minnesota County Radon Concentrations, 1987–88

County	Observed GM	Best estimate GM
Carlton	96	109
Cottonwood	97	120
Crow King	97	108
Clearwater	100	122
Martin	100	116
Swift	100	122
Norman	103	126
Wadena	103	121
Fillmore	105	131
Becker	107	128
Chisago	107	122
Isanti	107	128
Morrison	109	120
Sherburne	111	122
Ramsey	112	116
Kittson	115	131
McLeod	118	124
Beltrami	121	129
Yellow Medicine	122	136
Meeker	126	134
Olmstead	126	129
Marshall	127	132
Kanabec	128	136
Sibley	129	136
Benton	130	137
Roseau	131	134
Washington	131	132
Pope	134	140
Hennepin	136	136
Dakota	137	137
Rock	137	141
Carver	144	143
Otter Tail	144	143
Mahnomen	145	143
Polk	146	144
Stearns	148	147
Cass	151	147
Renville	156	148

Table 4-4 (*Continued*)

County	Observed GM	Best estimate GM
Winona	163	157
Todd	164	150
Lake Of The Woods	168	153
Big Stone	169	152
Houston	172	158
Scott	181	170
Steele	181	167
Wright	182	170
Mower	183	171
Le Sueur	185	163
Brown	189	162
Douglas	194	174
Pipestone	200	166
Wabasha	208	178
Chippewa	210	170
Rice	221	193
Clay	222	198
Stevens	222	162
Dodge	224	169
Traverse	231	177
Redwood	234	183
Goodhue	235	206
Lyon	242	198
Blue Earth	250	216
Nobles	255	178
Freeborn	259	210
Jackson	280	200
Kandiyohi	291	196
Lincoln	314	203
Nicollet	323	206
Watonwan	344	199
Wilkin	344	166
Murray	448	173
Lacquiparle	498	205
Grant	0	NA
Red Lake	0	NA

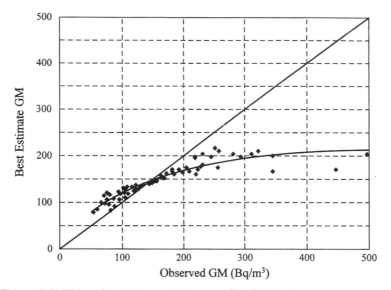

Figure 4-4. The estimated geometric mean (GM) of radon measurements for each county in Minnesota plotted against the actual GM taken from the set of measurements. Note the divergence of the estimated and actual measurement sets at both the low and high ends of the distribution.

question. For a nice set of environmental applications of Bayes' theorem, see *Ecological Applications*, volume 6, number 4 from 1996.

Monte Carlo Analysis

Monte Carlo analysis is another technique that is used when either the problem definition is unclear or the data available are uncertain. Monte Carlo analysis has become increasingly popular because it is such a natural application of computer resources in the area of risk analysis and probability assessment. Essentially, Monte Carlo analysis involves conducting and then comparing repeated trials with inputs that sample the distributions of the system parameters.

Problem 4-4 provides all of the needed distributions. In practice, finding the appropriate distributions can be the toughest part of the problem, especially as computer speed has reached the point that running a Monte Carlo program takes very little time.

Tap water in showers can be an important exposure route for a number of compounds that are absorbed through the skin, volatilized and inhaled, or both. The amount of time that people spend in the shower can have a direct effect on exposure levels; consequently, the distribution of shower lengths can be informative in assessing exposure to tap-water-borne pollutants.

We encourage anyone using this book to get a Monte Carlo software package (although the real diehards can write the code themselves). Some packages allow the user to explore distributions for which we have insufficient space to discuss. The Weibull, for example, which can represent a range of distributional shapes by varying three parameters, is used in problem 4-4.

Problem 4-4. Exposure to Tap Water in the Home

Given the data in table 4-5,

   a. what is the mean level of exposure of an "average" home occupant to tap water in the shower during the time he or she lives in a given house?
   b. what is a "conservative" value for the same number?
   c. what range of exposures can we expect to find for homeowners in the U.S. population as a whole?

Solution 4-4

Solution 4-4a

The mean value is a straightforward calculation. The total shower exposure (TSE) over ten years is equal to the water use rate times the

Table 4-5 Tap Water Exposure Parameters and Distributions

Parameter	Units	Mean	95th percentile	Distribution
Water use rate[a]	l/hr	480	800	Lognormal
Exposure time[a]	hours/day	0.13	0.28	Lognormal
Residence occupancy period[b]	years/life	13.2	41.4	Weibull

[a]After Finley and Paustenbach 1994.
[b]After Finley et al. 1994.

daily exposure time times the number of days in ten years, or

$$TSE_{mean} = 480 \ 1/hr \times 0.13 \ hr/day \times 365 \ days/year \times 13.2 \ years$$
$$= 0.30 \times 10^6 \ liters.$$

Solution 4-4b

A "conservative" calculation can be based on the assumed goal of protecting a maximally exposed individual. An approximation of this individual might use the 95th percentile value for each of the three parameters. Note that this is only one of many possible assumptions, and in later chapters we will show how these assumptions have been made in real and hypothetical cases.

$$TSE_{conservative} = 800 \ 1/hr \times 0.28 \ hr/day \times 365 \ days/year \times 41 \ years$$
$$= 3.4 \times 10^6 \ liters.$$

There is clearly a large difference—an order of magnitude—between the mean estimate and the "conservative" estimate. How realistic is either one? The former probably depicts the average homeowner fairly well, but provides no information about the range of individual exposures. The latter suggests that exposure might be quite high, but is based on the questionable assumption that the individual who is at the high end for one parameter is also at the high end for the other two. While this might be credible for this case, it is less realistic as the number of parameters increases, and can become absurd when, for example, the person who eats the most of one type of food also eats the most of all other types of food.

Solution 4-4c

It is also possible (but extremely time consuming) to solve mathematically for distributions of combined distributions. The Monte Carlo process approximates this solution by repeatedly using a *random number generator* to pick one value for each parameter based on that parameter's distribution and solving the equation based on those randomly selected numbers. Once many values have been found this way (generally at least 10,000 values), the resulting range represents fairly well the distribution of output values given the input distributions.

Figures 4-5 to 4-7 show the outcomes of 10,000 Monte Carlo runs using Crystal Ball for total shower exposure given the distributions listed above.

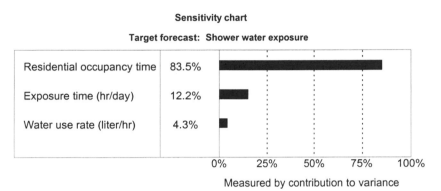

Figure 4-5. Sensitivity analysis of shower water exposure for problem 4-4c. This figure was generated by Crystal Ball, and is based on 10,000 Monte Carlo iterations of the model.

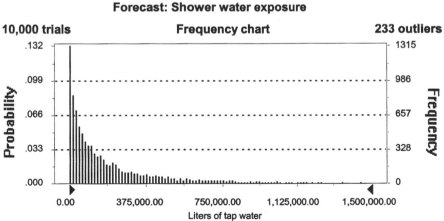

Figure 4-6. Probability distribution of predicted shower water exposure, for problem 4-4c. This figure was generated by Crystal Ball, and is based on 10,000 Monte Carlo iterations of the model.

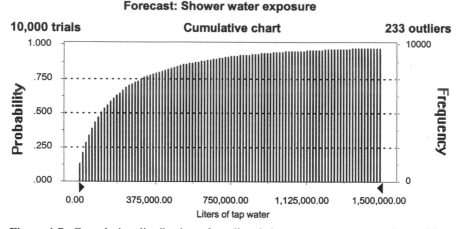

Figure 4-7. Cumulative distribution of predicted shower water exposure, for problem 4-4c. This figure was generated by Crystal Ball, and is based on 10,000 Monte Carlo iterations of the model.

Some Notes about Monte Carlo Methods

Burmaster and Anderson have proposed fourteen "principles of good practice" for using Monte Carlo techniques for risk assessment (Burmaster and Anderson 1994). The discussion of our example will illustrate how these principles are relevant in practice, although ours will be an abbreviated version. Note that Burmaster and Anderson suggest that "before an analyst undertakes a MC risk assessment...she or he will read widely in the growing literature on probabilistic risk assessment," a sentiment with which we strongly agree. Keep these principles in mind as you do the Monte Carlo problems in the rest of this book.

Principle 1: *Show All Formulae*

In our simple case, there is one main equation, that for TSE given above. All of the data used are shown in table 4-5, except as further developed below.

Principle 2: *Calculate Point Estimates*

We have calculated point estimates for mean and "conservative" exposures. Other, intermediate values could also be assessed.

Principle 3: *Perform a Sensitivity Analysis to Find Dominant Values and Distributions*

Since our output is a product of several parameters, there is no dominant value. However, a sensitivity analysis for dominant distribution shows that variation in residence occupancy period drives the final distribution, accounting for about 84% of the variance. If we are relatively uncertain about this distribution, we have cause for concern about the accuracy of our final output distribution. On the other hand, if the entire risk assessment included drinking tap water, the shower term itself might be relatively insignificant. The very small influence of water use rate on the variance suggests that a conservative or mean point estimate would have provided similar results. However, since a distribution for this parameter was readily available, we chose to include it.

Principle 4: *Only Do a Probablistic Risk Assessment (PRA) If Conservative Point Estimates Are Significant*

A properly executed PRA is highly resource intensive. Consequently, if a conservative point estimate indicates that risks are below some threshold, PRA is extraneous. Since we are not trying to find some minimum level, this principle does not apply to our example.

Principle 5: *Provide Detailed Input Information*

Table 4-5 above shows the basic information that we used for this example. More complete information is available in the two references. The distributions for water use rate and exposure time (ET) came directly from Finley and Paustenbach (1994). However, for the residential occupancy period (ROP), we used a curve that approximated the empirical data provided. Table 4-6 compares our distribution curve with the data from Finley et al. (1994).

Table 4-6 Empirically Derived Residential Occupancy Periods (Finley et al. 1994) Compared with the Authors' Estimates

	Percentile							
Source	*5th*	*10th*	*25th*	*50th*	*75th*	*90[tt]*	*95th*	*99th*
Finley et al.	0.28	0.56	1.4	5.2	17.1	32.0	41.4	64.2
Authors' estimates	0.28	0.67	2.32	6.92	16.5	31.6	43.55	55[a]

[a] Maximum value returned in 10,000 runs; 97.5th percentile.

The data were fitted through trial and error setting the Crystal Ball Weibull assumption to location 0, shape 0.8, and scale 11, and a maximum value of 100 years. In general, this appears to fit the Finley et al. data fairly well, with the 25th percentile value the farthest off. However, it is off by a factor of two, where the overall spread from the 5th to the 99th percentile is about two orders of magnitude. One point of concern may be the apparent truncation of the high end. A likely explanation is that this is simply the result of the fact that such long occupancies in a single home are quite rare. On the other hand, there may be something in the model that prohibits higher numbers from showing up.

Principle 6: *Show Input Variability and Uncertainty*

Finley et al. chose to express variability as 90th percentile/10th percentile. We do the same in table 4-7.

Uncertainty is a stickier issue. All three parameters are taken from studies of human behavior and are based on individual reporting, either of which can be distortionary. In addition, we are using past behavior to predict future behavior. In the short term, this is reasonable, but as the time frame expands it becomes increasingly problematic. Finley et al. describe the ROP as "moderately uncertain"; since we found under principle 3 that this is the "driving" statistic, it would be useful to look more closely before including it in a risk analysis.

Principle 7: *Use Real Data*

If there are no real data, generate them. If none are available, be very careful about how you model them (and document how you do it).

In our case, we used data from a refereed journal—two references from the journal *Risk Analysis*. In addition, we documented above how we transformed a data table for ROP into a distribution.

Principle 8: *Discuss Methods and Goodness of Fit*

This has been covered above to a large extent. Because we did not use the original numbers for ROP, a more rigorous interpretation of our Weibull curve is not possible.

Table 4-7 Ratio of Ninetieth to Tenth Percentiles for Three Input Parameters

Parameter	WUR	ET	ROP
Variation	2.3	4.6	47

Principle 9: *Discuss Correlation*

Burmaster and Anderson suggest that a "strong" correlation is $|\rho| \geq 0.6$. Neither of our sources reported any correlation among the numbers. It is possible to hypothesize a correlation between ET and WUR: if water pressure is low, people may take longer showers. On the other hand, it may be that people who take long showers also like high-pressure showers. It seems improbable that there is a significant correlation.

Principle 10: *Provide Detailed Output Statistics in Numerical and Graphical Form*

Table 4-8 shows the report generated by Crystal Ball on key statistics.

Principle 11: *Perform a Probabilistic Sensitivity Analysis for Key Inputs in Such a Way as to Clearly Distinguish between Variability and Uncertainty*

Part of this has been covered above, since we found that ROP is the key variable, as well as the most uncertain variable.

Principle 12: *Do Enough Runs to Ensure a Stable Output*

Burmaster and Anderson suggest at least 10,000 runs, which is what we did.

Table 4-8 Outputs from 10,000 Monte Carlo Runs

Statistics	Value	Percentile	Liters of tap water
Trials	10,000	5%	4,300
Mean	2.7×10^5	10%	10,000
Median	1.1×10^5	25%	3.7×10^4
Standard deviation	4.5×10^5	50%	1.2×10^5
Variance	2.0×10^{11}	75%	3.2×10^5
Skewness	5.8	90%	7.1×10^5
Kurtosis	79	95%	1.1×10^5
Coefficient of variability	1.7		
Range minimum	0.07		
Range maximum	1.2×10^7		
Range width	1.2×10^7		
Mean standard error	4,500		

Principle 13: *Use and Document a Sufficiently Random Random Number Generator*

A substandard algorithm or insufficient recurrence period will mean either bias or repeated samples. We used the default random number generator in Crystal Ball, which is good up to at least 10,000 runs.

Principle 14: *Explicitly Identify Limitations, Bias, and Further Study Needs*

This exposure model is probably reasonably well suited to an average group of homeowners. This is largely an artifact of the problem we selected: one with existing and generalized data. If this exposure assessment were to be used for a particular contaminated drinking water source, it would be important to assess how well these general data reflect the actual specific conditions.

The weakest point of our analysis is the residence occupancy distribution. A more detailed model of this distribution, based on empirical data, would be valuable. In epidemiological studies in general, this is often the most difficult part of the system to characterize with any certainty.

Problem 4-D. Uncertainty or Incommensurability?

Finkel (1995) argues that unacknowledged uncertainty in point estimates may be a more serious hurdle to risk comparisons than is the "incommensurability" issue. By incommensurability we mean that the data available do not provide a sufficient basis to compare the risks from different chemicals or risk pathways, or that the risks themselves are so different (e.g., the potential risks of skydiving versus the risks of cancer from pesticide residues on foods) that a common metric is often arbitrary. Finkel points out that claims, such as Ames and Gold (1989), that risk due to exposure to aflatoxin (a naturally occurring carcinogen in peanut butter) is "18 times worse" than that due to unsymmetrical dimethylhydrazine, or UDMH, are misleading. UDMH is a decay product of Alar, an infamous growth regulator that was once used frequently on apples used for juice.

Consider the following model for risk from ingestion of a carcinogen:

$$R = (A \times C \times \beta)/BW,$$

Table 4-9 Characteristics of the Probability Density Functions for the Input Variables

Variable	Units	Mean	5th %ile	95th %ile	Percentile location of the mean
Peanut butter consumption	g/day	11.38	2.00	31.86	66
Apple juice consumption	g/day	136.84	16.02	430.02	69
aflatoxin residue	$\mu g/g$	2.82	1.00	6.50	61
UDMH residue	$\mu g/g$	13.75	0.5	42.00	67
aflatoxin potency	kg-day/mg	17.5	4.02	28.33	61
UDMH potency	kg-day/mg	0.49	0.00	0.85	43

Source: After Finkel 1995.

where R is the risk associated with ingestion, A is the amount of the foodstuff eaten per day, C is the concentration of the contaminant in the foodstuff, β is the carcinogenic potency of the contaminant, and BW is the body weight of the individual.

a. Using the data in table 4-9, calculate the daily cancer risk faced each day by a 20 kg child who eats an average amount of both peanut butter. Do the same for a child who drinks an average amount of apple juice.
b. Compare the ratio of average risk from aflatoxin in peanut butter to the average risk from UDMH in apple juice.
c. Using the distributions in table 4-9, run 10,000 Monte Carlo simulations of this ratio. Compare this to the ratio as determined by the point estimates in (b). You will need to create hypothetical distributions based on the means, 5th and 95th percentiles, and percentile locations of the means. Discuss your results.
d. Discuss how additional calculations of this sort might influence interpretations of table 1-1.

References

Ames, B. N., and Gold, L. S. (1989). "Pesticides, risk and applesauce." *Science* 244, 755–57.

Burmaster, David E. and Anderson, Paul D. "Principles of good practice for the use of Monte Carlo techniques in human health and ecological risk assessments." *Risk Analysis* 14(4):447–81.

Finkel, A. M. (1995). "Toward less misleading comparisons of uncertain risks: the example of aflatoxin and Alar." *Environmental Health Perspectives* 103(4):376–85.

Finley, Brent and Paustenbach, Dennis. (1994). "The benefits of probabilistic exposure assessment: Three case studies involving contaminated air, water, and soil." *Risk Analysis* 14(1):53–73.

Finley, Brent, Proctor, Deborah, Scott, Paul, Harington, Natalie, Paustenbach, Dennis, and Price, Paul. (1994). "Recommended distributions for exposure factors frequently used in health risk assessment." *Risk Analysis* 14(4):533–53.

Goldemberg, J., Johansson, T. B., Reddy, A. K. N., and Williams, R. H. (1987). *Energy for a Sustainable World*. Washington, DC: World Resources Institute.

Gordon, Danielle A. and Kammen, D. M. (1996). "Uncertainty and overconfidence in time series forecasts: Application to the Standard & Poor's 500." *Applied Financial Economics*, 6:189–98.

Henrion, Max and Fischhoff, Baruch. (1986). "Assessing uncertainty in physical constants." *American Journal Physics* 54:791–98.

Morgan, M. Granger and Henrion, M. (1990). *Uncertainty: A Guide to Dealing with Uncertainty in Quantitative Risk and Policy Analysis*. New York, NY: Cambridge University Press.

Nordhaus, W. D. (1994). "Expert opinion on climate change." *American Scientist* 82:45–51.

Price, P. N., Nero, A. V., and Gelman, A. (1996). "Bayesian prediction of mean indoor radon concentrations for Minnesota counties." *Health Physics* 71(6):922–936.

Schmitt, S. A. (1969). *Measuring Uncertainty: An Elementary Introduction to Bayesian Statistics*. Reading, MA: Addison-Wesley Publishing Co.

Shlyakhter, Alexander I., Kammen, D. M., Broido, C. L., and Wilson, R. (1994). "Quantifying the credibility of energy projections from trends in past data." *Energy Policy* 22:119–30.

Williams, R. H. (1991). *Electrical Engineering Probability*. St. Paul, MN: West Publishing Co.

5

Toxicology

Introduction

In a review of the role of science in informing the U.S. EPA's most recent ozone standard, two scholars wrote, "the policy makers who support tighter standards cite the epidemiologists. Those who resist tighter standards cite the toxicologists" (Wilson and Anderson 1997). Presumably, toxicology, which gets information about hazards from laboratory subjects (usually animals) under controlled conditions, and epidemiology, which compares exposure and effect in the real world, should give the same answers. However, a plethora of assumptions accompanies each method, and it is these assumptions that generate different, and even contradictory, conclusions.

Rather than jumping right into the number crunching part of toxicological and epidemiological models, we first explore the science behind various dose-response models, and the assumptions that each requires the modeler to make. While a few paragraphs cannot make the reader an expert on disease causation, it is nonetheless important to understand why certain models are valid or could be used. There remains substantial and sometimes overwhelming uncertainty about many of the mechanisms underlying end points such as cancer, reproductive defects, and tissue death. Nonetheless, modeling can be highly informative and may generate clear policy responses. Many of the basic models of the transmission and evolution of illness also provide insights and guidance when constructing such models as the sequence of events leading to an accident or the problems of distinguishing simple correlation from causal effect.

Whether or not exposure to a particular chemical will cause disease in humans is an issue that stretches science to and beyond its current limits. Since many substances are clearly toxic, there is often societal pressure to treat toxic chemicals as such. This combination of a popular mandate to regulate risks and massive uncertainty about the existence and magnitude of those risks necessitates well-conceived and well-derived models.

Critical Assumptions for Modeling Disease

Knowing which key assumptions are made when modeling toxicological or epidemiological information to exposed populations enables one to acquire a feel for the potential value and limitations of the results. Three classes of assumptions will be reviewed in this section, those relevant to both toxicology and epidemiology and those specific to each of the two methods. The first section will focus on a significant, extensively researched unknown: the mechanisms of carcinogenesis. The reader is encouraged to think about analogous assumptions made when modeling other end points, since similar or greater uncertainty pervades other mechanisms, such as those of disease, financial planning, and energy demand. Thus, the models for each process are primarily used to gain qualitative understanding, not to make forecasts of exacting precision.

Mechanisms of Carcinogenesis

High- to Low-Dose Extrapolation

In chapter 1, we saw that a linear, nonthreshold model implies that it takes one "hit" to achieve a unitary effect, and the probability of getting a hit is proportional to the dose level. The corresponding low-dose extrapolation assumes a linear relationship between dose and response. While often useful as a first-cut description, a one-hit model is not consistent with the current understanding of carcinogenesis.

Cancer appears to come about as a several-step process, in which mutations at a number of gene sites, and not simply one point mutation, have to take place before a malignant tumor develops. This process reflects a basic level of error tolerance in the DNA encoding system. A cancerous cell is one that has gained the capacity to grow rapidly, while at the same time ignoring messages from surrounding cells to stop growing. The former is associated with oncogenes, and the latter with tumor suppressor genes, both of which are associated with a number of interacting growth factors.

Oncogenes can be viewed as the cellular equivalent of a gas pedal glued to the floor. A number of growth factors normally interact to stimulate a cell to begin to reproduce (promotion) and then undergo mitosis (progression). Oncogenes either replace one or more growth factors or influence components of their signaling pathways, and stimulate inappropriate promotion and progression (Aaronson 1993).

Tumor suppressor genes, on the other hand, normally have roles that promote apoptosis (cell death), which they can do either as negative growth factors providing an active signal or by blocking positive growth factors. Genetic mutations that cause the loss of these genes disable the cell's response to suppressor signals, and are necessary for the massive replication of carcinogenesis. A number of specific tumor suppressor genes have been identified (Weinberg 1993).

Unfortunately, while the mechanisms behind carcinogenesis are becoming clearer, they are, so far, only useful in a few highly specific cases where extensive work has been done. Anderson et al. (1993) argue that "in general, information on induction of specific oncogenes or suppressor genes is not yet certain enough to provide guidance to risk assessors." Similarly, information about thresholds is insufficient to apply them to most carcinogen models. Croy (1993) writes that "available evidence has been interpreted by some to support the inference that practical thresholds exist for carcinogens by nongenotoxic chemicals, but this proposition is not universally accepted."

A two-hit model has been proposed as a possible simplified description of carcinogenesis (Knudson and Strong 1972). The first hit involves the production of one or more oncogenes, while the second is the disabling of the tumor suppressor gene, with the ordering of the mutations not important. Problem 2-2 introduced this model qualitatively, and the mathematical version will be developed explicitly in problem 5-2. In order to use this (or any) model to extrapolate to low dose, however, one must assume not only that it is valid in general but also that *the processes at high doses are identical to those at low doses*.

Alternatively, a multistage model can be applied. An example of a three-stage model is *initiation*, a reversible process involving the creation of an oncogene and/or disabling of a tumor suppressor gene, *promotion*, where the mutated cell begin to proliferate, and *progression*, where one or more cancerous cells begin to metastasize, or spread through to other locations in the body. Figure 5-1 is a conceptual depiction of one-hit, two-hit, and two-stage carcinogenesis.

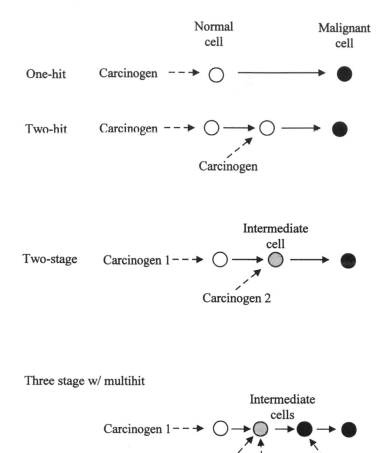

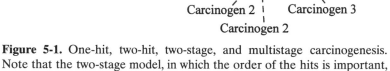

Figure 5-1. One-hit, two-hit, two-stage, and multistage carcinogenesis. Note that the two-stage model, in which the order of the hits is important, is a special case of the two-hit model.

It does not appear that all types of cancer follow the same pattern, nor that all carcinogens work the same way. Some carcinogens are not considered "genotoxic," which means that they do not initiate mutations, but may be active in the promotion or progression stages. Nongenotoxic promoters may have thresholds, while genotoxins may not. Such differences among cancers and carcinogens create additional uncertainty when modeling carcinogenesis.

There are reasons to believe that high-dose exposures will either overestimate or underestimate true carcinogenic potential. Ames and Gold (1991) suggested profound problems with assuming that high-dose effects are identical to those at low dose. Particularly problematic is the possibility that rapid cell growth is stimulated by damage to cells at high doses. Their conclusion was that high-dose effects are tautological (i.e., the experimental conditions cause the disease in a way that would not occur outside the laboratory) in the sense that because high doses are used, cell growth occurs. If this high-dose mechanism is not relevant at low doses, then high-dose findings are necessarily overestimates. This may be compounded if high doses overwhelm the cell's capacity for self-repair (Gold et al. 1992). Consequently, DNA lesions that would not lead to fatalities at low doses do have an effect at high doses.

Infante (1993) argues that while this may be true, the science is not yet sufficiently mature to make assumptions about cell proliferation. In addition, he sees four other factors that may lead high-dose tests to underestimate risk. First, what he refers to as the "wasted dose" problem arises because tumors are assumed to have come about at the time the lab animals are sacrificed, which undercounts earlier (and consequently lower-dose) tumor inception. Second, the twenty-four-month lifespan of test animals misses any tumors that would have appeared later given the same dose, and it is often the case that the long-latency illnesses are those of greatest regulatory concern. Third is the cumulative versus sporadic dose problem mentioned above. Fourth, only one chemical is tested at a time, which means that potential interactive effects with other chemicals are ignored.

Another problem with trying to apply conventional models is that the human population is not a homogeneous group when it comes to predisposition to cancer. Knudson and Strong (1972) found that some individuals are born with an inherited predisposition to a particular retinal malignancy. He hypothesized that these individuals are born with one of the hits necessary to cause the cancer. Consequently, only a single hit is needed to cause the malignancy in those individuals, and for them, a two-stage model is inappropriate.

Since Knudson's work, a number of genetic predispositions have been isolated, and it is likely that others will be found. In addition, other predispositions may make the population heterogeneous. For example, it has been known for some time that acquired immunodeficiency syndrome (AIDS) sufferers are more susceptible to certain types of cancers, apparently due in part to viruses that carry oncogenes or generate growth factors. Again, even if the mechanisms of carcinogenesis are understood, this type of heterogeneity challenges the use of any particular model.

It is useful to distinguish between two different types of heterogeneity in populations. The first is systematic heterogeneity, which can be identified and analyzed. The other is random heterogeneity, which cannot be anticipated ahead of time. Our understanding of genetic factors is shifting from "random" to "systematic" as genes associated with certain predispositions are identified.

Interactions among Chemicals

Knowing that cancer is often a multistep process, the question of how it may result from chemical interactions becomes a very important one. Unfortunately, the above complications are compounded when this multichemical problem is considered. There are very little data on how the chemicals that have been tested interact. Particularly troubling is the possibility that one chemical will promote one step, while another chemical promotes another, meaning that in combination they are much more potent than a single carcinogen alone. Perhaps ironically, it is this same issue that causes the opposite problem for epidemiology, when there are so many different exposures that it becomes very difficult to isolate a single one.

So Why Model?

At this point it might seem that modeling is a pointless exercise. Why bother to do it at all, if the uncertainty is so great? There are at least two reasons. First, recall that the point of modeling is not necessarily to predict effects. Rather, modeling provides information about possibilities, and about which assumptions are the most important. If very different models suggest roughly the same outcome, then the assumptions may not be that important. If they yield very different outcomes, then the assumptions will dominate, providing a starting point for further research.

Second, in the face of dramatic uncertainty, the range of credible models can provide useful information about what the possibilities are, and may suggest which alternatives are most viable. Five different models may provide cost estimates for a policy choice that range over several orders of magnitude, but the choice may still be quite easy if the highest of the associated costs is ten dollars for a life saved! Highly imperfect information, so long as it is recognized as such, can be far more useful than no information at all.

Assumptions Specific to Toxicology

Toxicology generally requires extrapolation from lab animals or tissue samples, exposed to constant high doses of a chemical (or other stimulus) over a long period of time, to humans exposed to lower and often varying concentrations, often via a different medium. Consequently, a number of assumptions must be made about high-to-low dose and species-to-species comparisons, exposure route comparisons, and interactions among multiple toxins. The previous section discussed some of the high-to-low dose assumptions for carcinogenesis; similar assumptions are made for noncancer endpoints. The following paragraphs will explore the other assumptions.

Animal and Human Comparisons

It has been known for a long time that metabolism varies significantly among mammals of different body weights, and it is likely that this difference is significant when it comes to processing toxins. Researchers such as Wilson and Crouch (1987) have found that, while the essential metabolic processes are similar, metabolism appears to vary as a function of body weight. Recent research has converged on [body weight]$^{\frac{3}{4}}$ as a plausible conversion factor for oral exposure (Watanabe et al. 1992), and the EPA has adopted this as its default assumption (U.S. EPA 1997). This calculation is specified in problem 5-2.

Another issue in comparing across species is whether the mechanisms and target organs or tissues are analogous. In some cases, a chemical may affect an organ in a test species that is not present in humans. Interpreting such data can be difficult: should one assume that there will be no effect in humans, or that there might be an effect in some organ that humans have but not the test animals? To date, this issue has been considered for only a few regulatory decisions, but it is possible that a better understanding of mechanisms will provide a better theoretical basis for making decisions.

Susceptibility

Animal tests are often done on either highly susceptible, homogenous animals or outbred strains of high genetic diversity. One consequence of using homogenous groups is that there are often very high cancer rates, even in the zero-dose "control" group. Figure 5-2a shows dose-response data from a test on 1,3-butadiene. It appears that there is some sort of sublinear slope. However, a couple of observations challenge the validity of a truly sublinear relationship. First, there is a high response in the zero-dose group. Second, since the number of test animals is small (usually a group of 50–100 at each dose level), each data point actually represents the midpoint of a range of possible values. Figure 5-2b depicts these data, including the uncertainty associated with the small number of animals.

Why are homogenous and highly susceptible animals tested in order to evaluate risks to humans? There are two reasons, one of which is practical, the other health conservative (note that "health conservative" is a value decision, clearly inseparable from the analysis). The practical issue is the need to insure a detectable response, which may be absent in a "normal" animal population, and the need to insure stability between the test groups. Underlying the health-conservative issue is the expectation that some humans may be more susceptible. The desire to insure that the effect on those susceptible individuals is captured leads to a preference for susceptible rats. It is important to note the extent to which unavoidable risk management decisions pervade these very "behind the scenes" aspects of risk assessment.

In contrast, tests on genetically heterogeneous groups of animals may result in lower tumor rates. The diversity, however, leads to a different type of uncertainty. While the intentionally susceptible group is arguably definable as "worst case," there is no equivalent classification for heterogeneous groups, since there is no useful way to determine how a heterogeneous animal population relates to a heterogeneous human population.

Constant versus Sporadic Doses

There is considerable evidence that, for many chemicals, it is not only cumulative exposure that affects potency, but also the way in which exposure occurs (Cox 1996). Since the doses received by lab animals are

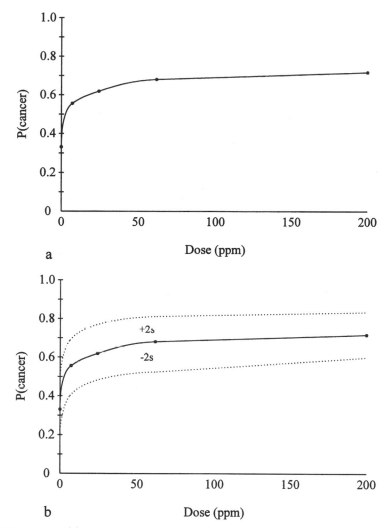

Figure 5-2. (a) At the sampled dose levels, the observed carcinogenic response to 1,3-butadiene appears to exhibit a concave pattern. Note, however, that the "baseline" in this sample was significantly elevated: over 30% of the zero-dose group rats had tumors. (b) The 95% percent confidence interval around the observed carcinogenic response to 1,3-butadiene exposure. The small sample size and high baseline render the expected dose-response relationship highly uncertain even within the observed dose range.

constant rather than sporadic, the assumption that a cumulative dose delivered at one rate is equivalent to doses at all other rates is problematic. For example, if there is a threshold, the effects could be dramatically different in two animals receiving the same cumulative lifetime dose, one at a constant daily rate below the threshold and the other in monthly doses, each of which is above the threshold. Consequently, it would be valuable to know what the highest levels of exposure will be, and how long they will last. Unfortunately, it is often very difficult to calculate cumulative exposure, much less the pattern of that exposure; thus the unsatisfactory but necessary assumption that cumulative exposures are the same, regardless of pattern.

Exposure Routes

There are three main routes by which toxins enter the body: the stomach, the lungs, and the skin. While the routes will be discussed in more detail in chapter 7, some consideration is useful here because animal tests are often done on one of these routes, and the results extrapolated to others. One possibility is to simply make a one-to-one comparison, in terms of total amount of exposure. Alternatively, a health-conservative assumption might be that the actual exposure route is more hazardous than was the experimental route, but it should be easy to see how such an argument becomes circular. One promising technique called "pharmacokinetics" (see problems 2-3 and 2-4) looks in detail at how a particular toxin travels in the body.

Interactions among Multiple Toxins

While an advantage of isolating an individual toxin is that others can be ruled out, there is evidence that for many toxins, combinations may have more than additive effects. For example, radon and cigarette smoking both clearly contribute to lung damage and in combination are much more potent than either alone (U.S. Department of Energy 1990). This association was uncovered through epidemiology studies and is unlikely to have been the subject of a toxicology test. Similarly, self-repair mechanisms that allow an organism to tolerate some amount of either of two toxins individually may be overwhelmed by the same amount of both at the same time. Testing the two independently might lead to standards that are less health protective than the combined effect would warrant.

Advantages of Toxicology Assessments

While the past several paragraphs suggest a number of limitations, toxicology remains an indispensable tool for risk analysis. Animal studies can indicate whether we should expect a toxin to have an effect on humans and suggest the relative potencies of different toxins. When information about mechanisms is available, valuable extrapolations can be made. Because it is prohibitively expensive to test large numbers of animals at low doses and usually unethical to test human subjects at any dose level, small animal samples provide relatively inexpensive screening data.

Assumptions Specific to Epidemiology

Epidemiology looks at actual exposed populations to evaluate the effects of various exposures. A study by John Snow in the 1800s linking cholera in London to drinking water supply provided insight into the cause and transmission of disease and helped define population studies, the basic study method for a range of issues, such as economics, crime, and health. Studies on workers under diverse workplace conditions provide strong data on the beneficial and harmful effects of exposures and practices. Epidemiological studies can include large numbers in both "test" and "control" samples, providing considerable statistical power to the results.

Two major types of epidemiological studies are "descriptive" and "analytic" (Gots 1993). Descriptive studies look at existing populations, isolate past exposures, and relate these to suspected effects. Frequently, these studies provide data on more extreme exposures than would be possible in analytic studies, which try to isolate a specific exposure in a forward-looking study. A major limitation of descriptive studies is that of estimating historical exposures and other conditions. Double-blind tests for new medicines are well-known examples of analytic studies.

Epidemiology faces a different, but no less vexing, set of complications than does toxicology. In general, the populations that are analyzed are exposed at higher doses than are other populations of interests, so dose-response questions don't go away. Other key problems include isolating confounding factors, imperfect measurements, representativeness of the analyzed population, and interpreting small effects.

Confounding Factors

Confounding factors arise because epidemiological experiments are "natural" rather than "controlled." It is often the case that a group of individuals that are exposed to one toxin are also exposed to other stresses as well. It may be clear that there is an elevated rate of disease, yet still not obvious which or what combination of a number of factors is the cause. Well-developed statistical tools can be a significant help in this area, but do not always provide answers.

Measurement Error

Another consequence of "natural" experiments is that measurements of exposures are reconstructed and are nearly always imperfect. Ideally, complete information on exposure levels and rates for each member of the analyzed group can be found, but generally, average levels must be used. If no exposure records were kept, then ventilation, duration of exposure, and emission or ingestion rates would all have to be estimated. Even when records are available, they may be incomplete or otherwise unreliable.

Representativeness

The extent to which the analyzed group is similar to the general population can have important implications for extrapolation. One frequent example is the comparison of exposed workers to the general population. Workers may be more healthy or have other particular characteristics that differentiate them from the general population. Identifying the differences, and the potential implications of the differences, can be a challenge.

Interpreting Small Effects

Relative risk (RR) is a standard epidemiological unit (see problem 2-5). A RR of 1 means that exposure does not create any additional risk, while a RR of 2 means that exposure leads to twice the risk of nonexposure. Studies show that the RR of lung cancer for an American male smoking twenty cigarettes per day is about 40; there is little (credible) debate about the validity of this figure. However, many epidemiologists are reluctant to support decisions based on values of RR < 2. Even when they feel that the above problems have been isolated and that the sample population is large enough to make the numbers statistically significant, the true uncertainty that remains may preclude a clear finding. It is not unusual for two epidemiologists to agree on the "science" behind a particular study, yet disagree on how it should be incorporated into the policy process.

Cause and Effect

Finally, no matter how good the data, epidemiological studies cannot provide clear information on cause. Even in a study where 100% of controls are unaffected and 100% of test individuals are, it is impossible to say what about the given exposure is causing the effect, only that the effect happens. Consequently, combining toxicological studies, which can focus much more specifically on mechanisms, with epidemiological studies can provide more complete information than either type of study alone.

This chapter and chapter 6 contain a range of toxicological and epidemiological problems. The solutions try to address the abbreviated list of critical assumptions above, but in the interest of space, it is impossible to address them all. The reader should keep in mind that many assumptions have to be made, and think critically about each step. Instructors using this book are encouraged to assign not only the unsolved problems but also further exploration of the problems we have presented as "solved."

This chapter examines how toxicological data can be used to estimate the possible human health effects of toxins. The first problem uses a conveniently linear, hypothetical data set to explore the kinds of conclusions we can postulate about carcinogens, and the assumptions we must make to model relationships among doses, species, individuals, and end points.

Problem 5-1. Test Data and Carcinogenesis: The Kil-EZ Example

Consider the following hypothetical situation. Kil-EZ, a pesticide, is used to protect fruit trees from several insects, and when applied normally to these crops, it is found in harvested fruit. A regulatory agency believes that Kil-EZ might be a human carcinogen. Consequently, they require the manufacturer to test Kil-EZ on a special strain of laboratory rats, which are chosen because they are particularly susceptible to carcinogenesis. The test requires that the rats be given Kil-EZ via oral gavage (delivered through a tube into the stomach), at several dose levels up to the point at which they exhibit acute symptoms, such as loss of body weight. The test comes back with the data in table 5-1. Based on these data,

  a. is it reasonable to conclude that Kil-EZ is a rat carcinogen?
  b. is it reasonable to conclude that Kil-EZ is a human carcinogen?
  c. if Kil-EZ is a human carcinogen, how potent is it?
  d. should Kil-EZ be regulated to avoid human cancers?

Solution 5-1

This question reflects the issues facing a number of agencies in the United States and abroad during the process of making decisions about chemicals in the environment. To answer it, we need to examine the

Table 5-1 Hypothetical Data for Kil-EZ Carcinogenicity Test

Number of test rats	Number of rats with tumors	Rat test dose in mg Kil-EZ per kg rat body weight per day (mg / kg / d)	Exposure
50	1	0	NO
50	5	20	YES
50	10	40	YES
50	20	80	YES
50	7[a]	160	YES

[a]Acute effects were observed in the 160 mg/kg/d group, leading to significant premature deaths.

theory and the data to decide which models to use for the relationship between high doses and low doses of a single chemical, and for the relationship between animal and human responses.

It may seem dubious to make policy decisions based on data as limited as those in table 5-1, but this is a fairly realistic case. The real data would probably have more details on the types of cancers found, but the number of rats with tumors is generally the driving datum for toxicological risk assessment. The reason why such limited data are considered acceptable is illuminated by the cost of such a study—in the vicinity of one to five million dollars.

Solution 5-1a

A quick look at the available data suggests that the answer to this question is likely to be yes. The key assumption is that the animals in all groups were treated identically except for Kil-EZ exposure level, which allows the inference that any difference in cancer frequency between the groups is caused by the Kil-EZ. While it appears clear from the lack of results in the zero-dose group that Kil-EZ is carcinogenic (although there may be some effects here, too, due to the fragile nature of laboratory rats), a formal statistical test is appropriate. Specifically, the aim is to test whether there is or is not a difference between the zero-dose group and the nonzero-dose groups, and for each rat, there are two possibilities: either it grows tumors or it does not. Consequently, this test can be treated as binary (tumor versus nontumor):

- Assign the value 0 to tumor-free rats and the value 1 to rats with tumors
- Define p as the probability that a rat will grow tumors, which is the number of rats with tumors divided by the total number of rats in the sample
- Recall from chapter 3 that the expected (mean) value of a binomial function is the number in the sample times the probability of an effect

For the zero-dose case, the estimated probability of tumors is

$$P(\text{tumor} \mid \text{zero dose}) = (1/50) = 0.020,$$

and the expected number of rats with tumors is

$$E(\text{tumor} \mid \text{zero dose}) = n \times p = 50 \times 0.02 = 1.$$

For the nonzero-dose group, there are 35 rats with tumors out of 150 total tested, so the probability of an exposed rat growing tumors is

$$P(\text{tumor} \mid \text{nonzero dose}) = (35/150) = 0.23.$$

We can now test whether the exposed and nonexposed groups are the same. If they are the same, then

- the mean of the differences between the proportions is 0, and
- the standard deviation of the differences between the proportions is

$$\sqrt{Q \times (1 - Q)\left(\frac{1}{n(\text{zero})} + \frac{1}{n(\text{nonzero})}\right)},$$

where

$$Q = [E(\text{zero}) + E(\text{nonzero})]/[n(\text{zero}) + n(\text{nonzero})],$$

or

$$s(\text{difference}) = \sqrt{0.18 \times (1 - 0.18)\left(\frac{1}{50} + \frac{1}{150}\right)} = 0.063.$$

Using these results, we can calculate a Z-score,

$$Z = \frac{P(\text{zero}) - P(\text{nonzero})}{s(\text{difference})} = \frac{0.02 - 0.23}{0.063} = -3.33.$$

Choosing 99% as the desired confidence level, the Z-table in appendix A gives the Z-value for 0.01 as 2.33. We can therefore reject the hypothesis that the zero- and nonzero-dose groups are the same.

The reader should also test whether there is a statistically significant difference between the 20 mg/kg/day group alone and the zero-dose group.

Solution 5-1b

This question addresses the comparability of a susceptible rat strain and humans. The conservative regulatory assumption is that, unless there are factors that suggest that the particular tumors observed are specific to the test animal, carcinogenicity in any animal implies carcinogenicity in humans. The EPA calls this assumption a "science policy decision" (U.S. EPA 1997). There are both empirical and theoretical reasons for this assumption, but additional research in this area would be very valuable (National Research Council 1993). Most human carcinogens have also been found to be carcinogenic to other mammals (Infante 1993).

The fact that the animals were exposed orally supports the inference that Kil-EZ is a human carcinogen, since the primary human exposure to Kil-EZ is also oral. The fact that there is only one test, and no available human exposure data, however, might make us cautious about the comparison. Several other factors might make a difference if we knew them, such as whether the rats were all the same sex or a mix of sexes, and the type and severity of the observed tumors (National Research Council 1993).

The EPA draft guidelines classify carcinogens in three descriptive categories, "known/likely," "cannot be determined," and "not likely" (U.S. EPA 1997). Based on the data in table 5-1 alone, Kil-EZ would be placed in the "cannot be determined" category for human carcinogens. "Not likely" is inappropriate because the substance appears to be carcinogenic to some mammals, while "known/likely" is reserved for substances with more complete data, generally including epidemiological tests, toxicological tests on tissue samples, and so on.

Solution 5-1c

So far, this approach has concluded that Kil-EZ is a rodent carcinogen and may be a human carcinogen. But we would like to know more; we would like to understand the dose-response relationship. In the study, a homogeneous group of rats were fed a large and constant amount over their entire lifetimes, and now we want to infer the cancer risk for a heterogeneous human population eating small amounts sporadically. To do this, we must either find out or make assumptions about the following:

- The relationship between human metabolism and rat metabolism
- The relationship between lifetime exposures for rats and humans
- The relationship between a homogeneous susceptible group and a potentially heterogeneous group
- The relationship between constant doses and sporadic doses
- The relationship between high doses and low doses

Based on the discussion of the science behind these issues, a qualitative approach is an appropriate starting point. The current regulatory philosophy is that, while understanding has improved considerably over the last several decades, relationships among species are not well enough understood to make decisions about carcinogenic potency based on animal studies alone. The current EPA regulatory standard is much like the following qualitative approach, with a few modifications (U.S. EPA 1997).

First, without a known model of pathway differences, assume that humans and animals metabolize similarly, and simply adjust one-to-one for the difference in body weight. Doing this, one might expect that 20 mg_{kil-EZ}/kg(rat) is equivalent to 20 mg_{Kil-EZ}/kg(human). Later, we will examine some data that indicate this straight scaling is inappropriate, but for now the one-to-one ratio will serve.

Since laboratory rats typically live about three years, and humans about seventy, one must make assumptions about this relationship. The typical assumption is that daily exposures can be compared from species to species. Combining the body weight and lifetime assumptions, we find that a 100 kg human consuming 2000 mg of Kil-EZ per day would have the same cancer risk as a 0.5 kg rat eating 10 mg of Kil-EZ per day.

Since it is not clear whether sporadic doses will have a larger or smaller effect than continuous doses, a working assumption might be that they are essentially the same when accumulated over a lifetime. This assumption implies, for example, that 100 mg eaten all in a single day will have the same carcinogenic effect as 1 mg per day every third day for a little less than a year.

A decision about the relationship between high and low doses requires us to know something about the mechanisms of carcinogenesis So far, we assume that we know nothing. Consequently, a first assumption might be that the relationship is linear (as in the one-hit model

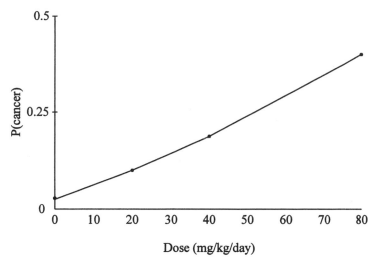

Figure 5-3. Observed carcinogenic response to the hypothetical Kil-EZ data of problem 5-1.

from chapter 2), and that there is no threshold level (since our observation of no response is seen only at the zero-dose level, and all nonzero doses have positive effects). These two assumptions (linearity and no threshold) suggest a relationship like that in figure 5-3.

The relationship between high and low doses can be represented by the equation for a straight line, or

$$p(\text{cancer}) = 0.1/(20 \text{ mg/kg/day}) \times \text{dose}.$$

From this, the risk of cancer at any dose can be calculated. For example, some regulations call for risk to be controlled to the one in a million (10^{-6}) level. This corresponds to

$$10^{-6} = 0.1/(20 \text{ mg/kg/day}) \times \text{dose}.$$

Rearranging,

$$\text{dose} = 10^{-6}/[0.1/(20 \text{ mg/kg/day})]$$
$$= 0.2 \ \mu\text{g/kg/day}.$$

This provides a rough idea of what the dose-response relationship might be, and allows us to poke at it a little to see what might be the more problematic assumptions. First, think about variability even given the assumptions. When extrapolating from the 20 mg/kg/day level, a 95% confidence interval can be calculated using the binomial, as in part a. The expected value of the number of affected rats given 20 mg/kg/day is 5, corresponding to $p(\text{cancer}) = 0.10$. The standard deviation for this value is

$$s = \sqrt{p(1-p)/\sqrt{n}}$$

$$= \sqrt{0.1 \times (1 - 0.1)/\sqrt{50}}$$

$$= 0.042.$$

The same calculation can be done for each of the other dose levels.

Using this measure of uncertainty, it is easy to see the effect on the extrapolation: adding and subtracting twice the standard deviation ($2s$) yields a $\{0.06, 0.14\}$ range of cancer risks associated with 20 mg/kg/day. The corresponding range of doses associated with a one in a million risk is $\{0.33~\mu g/kg/day,~0.14~\mu g/kg/day\}$! This is shown graphically in figure 5-4. The variability alone generates an order of magnitude of uncertainty about the extrapolation.

Most of the EPA's current carcinogenesis studies are based on the "linearized multistage" (LMS) model, and results of the studies that have already been completed based on this model will continue to dominate the decision-making process for some years to come. How-

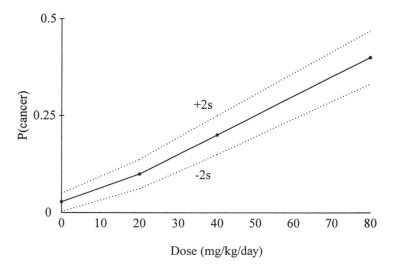

Figure 5-4. The 95% confidence interval around the observed carcinogenic response to the hypothetical Kil-EZ data of problem 5-1.

ever, as discussed in chapter 1, the EPA may abandon the LMS model in favor of a model-free curve-fitting and extrapolation technique similar to that used by the U.S. Food and Drug Administration (FDA). Figure 5-5 represents the EPA's proposed model, which is based on the dose associated with an excess risk of 10%, or ED10. This technique requires fitting a curve (via a polynomial or other model) to the available data and, through background tumor rates, finding the conservative 95% confidence interval for that curve, and identifying the dose associated with a 10% increase in risk. This dose is called the LED10, or lowest excess dose (defined as the lower 95% confidence level) associated with a 10% increase in risk. A straight line from the LED10 to the origin is used to compute the doses associated with various levels of risk.

This LED10 extrapolation represents the state of the art in extrapolation from an animal study without additional information. This may be

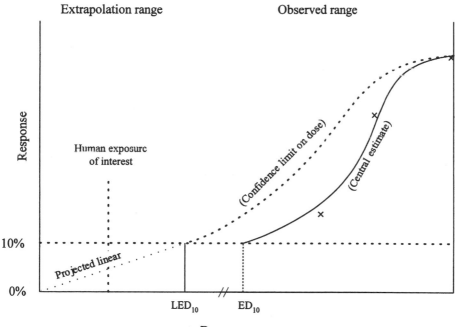

Figure 5-5. Graphical presentation of data and extrapolations. From U.S. EPA 1997, figure 3.1. Note that the data points from which the central estimate and 95% upper confidence limit are derived are located above the ED_{10}. The actual human exposures of interest are significantly lower than any of the test doses.

somewhat disconcerting, since it may not be much more meaningful than drawing a line from the lowest data point with a wide-point marker, and using the upper part of the line as the confidence interval. Nonetheless, it might be considered a reasonable assumption, "plausibly conservative" in regulatory terminology. Clearly, a better understanding of carcinogenesis would be extremely valuable.

A frequent observation is that LED10 extrapolation, mechanistic models, and nonmechanistic curve-fitting techniques (some of which are discussed in more detail below) all seem to produce similar low-dose risk estimates. An important conclusion is that improved understanding of the mechanisms of carcinogenesis is a potentially valuable research direction. Unfortunately, while rat assays are very expensive (the hypothetical test above would probably cost around 1–5 million dollars), research into the actual mechanisms can be more expensive still.

Problem 5-A. Calculating LED10

Calculate the LED10 for Kil-EZ, and use it to find the daily dose associated with a 1 in 100,000 risk and a 1 in 1,000,000 risk.

Problem 5-B. Maximum Tolerated Dose

The maximum tolerated dose (MTD), or the highest dose at which significant acute effects such as body weight loss are not observed, has been found to be correlated with carcinogenic potency. Krewsky et al. (1993) observe that, based on a large number of comparisons, a dose associated with a 10^{-6} risk can be estimated by dividing the MTD by 380,000. Using this factor, estimate the dose of Kil-EZ associated with a 10^{-5} cancer risk. As always, include both quantitative and qualitative measures of uncertainty.

Problem 5-C. 1,3-Butadiene

Recreate figure 5-3, using the 1,3-butadiene data in table 5-2.

Table 5-2 1,3-Butadiene Data

Number of test rats	Number of rats with tumors	Rat test dose (mg / kg / d)
70	8	0
70	26	6.25
70	33	20
70	41	62.5
70	48	200

Problem 5-2. Fitting Data to Mechanistic Models: One-Hit, Two-Hit, Two-Stage Problem

Table 5-3 contains the results of an animal experiment for determining the dose-response behavior of benzene (Crump and Allen 1984).

   a. Find the equivalent human dose for each mouse test dose.

   b. Fit a one-hit (i.e., linear), a multihit (two-hit), and a multistage (two-stage) model to the data set, and calculate the daily dose associated with a 10^{-6} excess cancer risk.

   c. Calculate the excess cancer risk associated with 1 ppm exposure, 40 hours per week, 50 weeks per year, for 30 years.

In each case, state explicitly any assumptions that you make. Comment on your results.

Table 5-3 Benzene Rat Test Data Set

Number of test mice	Number of mice with tumors	Mouse test dose (mg / kg / d) (oral gavage)
50	0	0
50	4	25
50	20	50
50	37	100

Source: Crump and Allen 1984.

Solution 5-2

A useful first step in solving this type of problem is to graph the data, to provide some idea of what might be going on. Figure 5-6 is a graph of the data in table 5-3. The dotted lines represent two standard deviations on either side of the data, or about a 95% confidence interval.

This problem covers many of the main steps of toxicological risk assessment, and is typical of the way that risk assessment is performed by regulatory agencies. The goal of toxicological risk assessment is to use data from small samples of laboratory animals exposed to high levels of a suspected carcinogen to model the effect that low doses of the same carcinogen will have on humans. The main (and highly contentious) assumptions in any such model are

- How laboratory animal doses correlate to human doses (scaling factor)
- How the effects from high doses compare to those from low doses (extrapolation)

Additional assumptions that are receiving more scrutiny as modeling becomes increasingly complex include

- How well short-term, high-dose exposures correlate to long-term, low-dose exposures (dose metric)

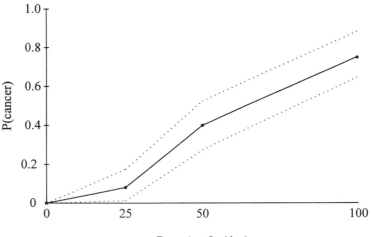

Figure 5-6. Graph of benzene data. The dotted lines represent a 95% confidence interval around the data. From Crump and Allen 1984.

- What happens to the doses inside the body (physiologically based pharmacokinetics, or PBPK)
- How doses from one medium compare with those from another medium (inhalation, ingestion, and dermal absorption)
- How much uncertainty is associated with small experimental sample sizes

Solution 5-2a

Dose Scaling. Because toxicological experiments are done on animals (typically mice and rats), it is important to try to understand the relationship between the dose the test animal receives and that received by a human when exposed to the same concentration. In the basic model, the *dose metric* is in terms of a total lifetime dose, or daily dose averaged over a lifetime. The *scaling factor* from test animal dose to equivalent human dose is a function of several factors, most importantly lifespan and body weight (BW).

The scaling factor equation is

$$D_{human} = sf \times D_{animal}, \qquad 5\text{-}1$$

where D_{human} is the lifetime average daily dose to human (in mg/kg body weight/day), D_{animal} is the lifetime average daily dose to animal (in mg/kg body weight/day), and sf is the scaling factor.

$$sf = (BW_{human}/BW_{animal})^{1-b},$$

where $b = 0.75$ is the default value recommended by the U.S. EPA (1997). Watanabe et al. (1992) find that $b = 0.74$ fits the available data, but argue that it is not clear that there is a single scaling factor valid in all cases.

Average values for body weights and lifespans are found in table 5-4.

Using the formula for dose scaling and the data above we can fill in the corresponding equivalent human lifetime doses (table 5-5). It is assumed that daily doses over a lifetime are equivalent, regardless of differences in species lifetimes.

$$D_{human} = (71/0.03)^{0.25} \times D_{mouse}$$
$$= 7.0 \times D_{mouse}.$$

Table 5-4 Human and Rat Mean Body Weights and Lifespans, by Gender

Species	Sex	Standard lifespan value (years)[a]	Standard adult BW (kg)[a, b]	Adult BW distribution[b]
Human	Male	72	78.7	Normal (78.7, 13.5)
	Female	79	65.4	Normal (65.4, 15.3)
	Both	76	71.0	Normal (71.0, 15.9)
Rat	Male	2.5	0.5	NA
	Female	2.5	0.35	NA
Mouse	Male	2.5	0.03	NA
	Female	2.5	0.25	NA

[a] Hallenbeck 1993.
[b] Finley et al. 1994.

Table 5-5 Benzene Data Set with Equivalent Human Dose

Number of test mice	Number of mice with tumors	Mouse test dose (mg / kg / day) (oral gavage)	Equivalent human dose (mg / kg / day)
50	0	0	0
50	4	25	100
50	20	50	200
50	37	100	400

Solution 5-2b

So far we have established, with certain assumptions about the similarities between laboratory animals and humans, a relationship between the number of excess tumors we would expect to find in a group of fifty people exposed to each of four doses of benzene (table 5-6). We interpret this fraction to be the risk (probability of getting at least one tumor) for each person so exposed, LCP(D), which is the lifetime cumulative probability of at least one tumor at dose D. [Note: since this dose is the daily dose over a lifetime, time is implicit in the dose, which we would otherwise write $D(t)$.]

Noting that these data are for doses within a limited range (less than a factor of ten from 175 mg/kg/day to 700 mg/kg/day), the next task is to relate these high probabilities of tumor responses at quite high

Table 5-6 Expected Lifetime Probability of Death (LPD) for
Humans at Equivalent Human Dose (D)

LPD	D (mg / kg / d)
0	0
0.08	100
0.40	200
0.74	400

concentration levels to the probability of tumor responses at doses
several orders of magnitude below this. A variety of models have been
suggested and used, and these models can yield highly divergent results.
The models in use range from very simple models that assume a direct
linear relationship between high and low doses to highly complex
models that try to account for how individual cells become malignant
(hit, multihit, and multistage models), the extent to which the body can
tolerate a toxin without showing an effect (threshold models), how
chemicals are distributed and metabolized in the body (physiologically
based pharmacokinetic models), as well as what available epidemiological data suggest to us about real conditions.

One consequence of having so many different models is that the
uncertainty involved in comparing one model to the next may overwhelm the variability that one finds by using statistical methods to
evaluate a given model. Not surprisingly, major disagreements exist
about which models are the most realistic and, therefore, appropriate,
and the extent to which and ways in which margins of safety should be
incorporated into models.

One-Hit (Linear) Model. The one-hit model of carcinogenesis assumes
that a single event (or "hit") causes a cell to become cancerous. Based
on this assumption, lower concentrations mean that there is a propor-
tionally lower probability that a hit will occur. This suggests a linear
relationship that holds from high dose to low dose. Given the additional
assumption of lifetime exposure at constant dose, the one-hit model is
described by the following formula:

$$\text{LCP}(D) = 1 - \exp[-(q \cdot D)]. \qquad \text{5-2}$$

We can calculate the likelihood using the binomial distribution, as follows. Since for benzene we know three LCP(D)s for three high doses, equation 5-2 can be used to calculate q^* (an estimate of the parameter q). Because most models will not fit the data exactly, we look for parameter estimates that "maximize the likelihood" that the model fits the data. A model with estimated parameters yields an estimate $P^*(D_i)$ for each observed point $P(D_i)$. Consequently, to find a best fit for a model, we want to insure (i.e., maximize the likelihood that) the estimated data points as a group fit the actual data points as well as possible.

To do this, we find the likelihood function for any given $P^*(D_i)$ using information we know about the binomial distribution defined by the actual data point. We find that

$$\text{likelihood of } P^*(D_i) \text{ given } P(D_i)$$
$$= P^*(D_i)^{P(D_i) \cdot n} \times [1 - P^*(D_i)]^{n[1 - P(D_i)]},$$

where n is the number of test animals at the given dose. Intuitively, as $P^*(D_i)$ gets farther away from $P(D_i)$, it becomes increasingly unlikely that $P^*(D_i)$ reasonably represents $P(D_i)$.

Because the goal is to maximize all of the likelihoods, we multiply them together to find the likelihood that the model fits the data set, or

$$\text{likelihood of the model given the data set}$$
$$= \prod_i P^*(D_i)^{P(D_i) \cdot n} \times [1 - P^*(D_i)]^{n[1 - P(D_i)]}.$$

Since the numbers get very large or very small, it is sometimes useful to maximize the log-likelihood, where

$$\text{log-likelihood} = \sum_i \ln\left[P^*(D_i)^{P(D_i) \cdot n} \times [1 - P^*(D_i)]^{n[1 - P(D_i)]} \right].$$

In order to find the parameters that best fit the model to the data, we try to maximize the value of this log-likelihood, through what is called a bootstrapping technique. This means choosing initial values for the parameters, calculating the likelihood, and then choosing a second set of parameters a small step away and doing the same. By picking the better of the two resulting likelihoods and repeating the process, eventually, a set of parameters is obtained that cannot be improved; this is the maximum-likelihood estimate.

There are two problems with the bootstrapping method. First, the computations can take a lot of time, especially when there are many parameters (nonlinearity). Second, it is possible to find a "local" maximum or minimum, for which much better parameters can be found but not by moving a small step away. Computer programs are useful in overcoming the first of these, but only understanding the model equation and selecting appropriate initial (or "seed") values can address the second.

A program such as solver.xls with Excel will solve the one-hit problem readily, and the method is summarized below.

Step one:

Set up a spreadsheet with five columns (see table 5-7), one for animals, one for D, one for $P(D)$, one for $P^*(D)$, and one for log-likelihood. In addition, there must be two other cells, one which is the sum of column D, and the other which is q^*. It is usually necessary to insert an arbitrary "seed value" for q^*. (Try 0.001. Note that for the linear equation, it is almost certain that the q^* found by this method will be a "true" value, while for more complex equations, the choice of seed value may lead to a local minimum or maximum.)

Step two:

Use solver.xls or a similar program to maximize the value of cell $E6$ by changing the value of $C8$.

The maximum likelihood method yields

$$q^* = 0.0015.$$

For low-dose extrapolation, it is useful to realize that because $\exp(-x)$ $\approx 1 - x$ when $x \ll 1$, when $q \cdot D \ll 1$, equation 5.2 simplifies to

$$\mathrm{LCP}(D) = q^* \cdot D, \qquad\qquad 5\text{-}3$$

which is a linear dependence.

Equation 5-3 can be used to compute cancer probabilities at low doses. Using this model to find the dose associated with a one in a

Table 5-7 Representation of Spreadsheet for Computing the One-Hit Model

	A	B	C	D	E
1	No. of animals	Dose	P(D)	P*(D)	Likelihood
2	50	0.01[a]	0.01[a]	1 – exp – (B2 × C8)	ln(D2(C2 × A2) × (1 – D2)(A2 × (1 – C2)))
3	50	100	0.08	1 – exp – (B3 × C8)	ln(D3(C3 × A3) × (1 – D3)(A3 × (1 – C3)))
4	50	200	0.40	1 – exp – (B4 × C8)	ln(D4(C4 × A4) × (1 – D4)(A4 × (1 – C4)))
5	50	400	0.74	1 – exp – (B5 × C8)	ln(D5(C5 × A5) × (1 – D5)(A5 × (1 – C5)))
6				sum of likelihoods =	SUM (E2:E5)
7					
8		q^* =	0.001		

[a] Because ln(0) is undefined, we use an arbitrarily small nonzero term in the zero column.

million lifetime cancer probability,

$$10^{-6} = 0.0015 \times D,$$

so

$$D = 10^{-6}/0.0015$$
$$= 6.7 \times 10^{-4} \text{ mg/kg/day.}$$

Keeping in mind that there are substantial model uncertainties involved in extrapolating beyond the lowest data point, it is nonetheless instructive to think about the statistical variance, assuming that the model is correct. As was argued in chapter 3, the number of points in a sample has a strong influence on the credibility calculations. In addition, the costs of an animal study pressure researchers toward using the smallest sample size that will yield a statistically significant result.

Often, a health-conservative estimate of the dose associated with an increase in cancer is preferred to the most likely estimate. One way to do this is to use the value associated with the upper end of the 95% confidence interval. This q^* can be calculated using the method above, substituting the 95% $P(D)$ into column C. Note that this is not mathematically a 95% confidence interval on q^*, since we are combining 95% points for each dose level, and one would not expect the 95% confidence interval for the overall data set to equal a combination of the individual points. Problem 5-I explores a Monte Carlo approach to finding a 95% confidence interval for the whole data set.

Multihit Model. The multihit model is based upon the idea that more than one event is necessary to cause a cell to become cancerous. Consequently, $P_e(t)$ can be written as a sum of one-hit models. The following math appears complicated, but hold on—it simplifies in the end! For small increments,

$$dP_0(t)/dt = -\lambda(t)P_0(t),$$
$$dP_n(t)/dt = \lambda(t)P_{n-1} - \lambda(t)P_n,$$

where n is a positive integer number of hits, $P_n(t)$ is the probability that a given cell will contain n hits at time t, and $\lambda(t)$ is the transition rate (Crump 1981).

Since the value of each event at each time period is either 0 or 1, this becomes a Poisson process, so the probability that a given number of hits will occur can be written as

$$P_n(t) = [\Theta^n(t)/n!] \cdot \exp - \{\Theta(t)\},$$

where $\Theta(t)$ is an empirically derived term for the dose and cancer potency as a function of time. (See below for specific examples of $\Theta(t)$.)

If it takes k hits to transform a cell, then the probability that at least one tumor will arise is equal to the sum of probabilities that k or more hits occur, or

$$\text{LCP} = \sum_{n=k}^{\infty} P_n(t)$$

$$= \sum_{n=k}^{\infty} [\Theta^n(t)/n!]\exp\{-\Theta(t)\}.$$

Since the probability that each different number of hits occurs must sum to 1,

$$\text{LCP} = 1 - \sum_{n=0}^{k-1} P_n(t)$$

$$= 1 - \sum_{n=0}^{k-1} [\Theta(t)/n!]\exp\{-\Theta^n(t)\}.$$

Mercifully, assumptions of low dose and lifetime dose allow us to simplify. Assume a lifetime dose, so that $\lambda \times$ lifetime $\approx qD$ (because animal test studies use a constant dose over an entire lifetime). $\Theta(t)$ is approximated by $(q \cdot D)$, so

$$\text{LCP} = 1 - \sum_{n=0}^{k-1} \left[(qD)^n/n!\right]\exp(-qD).$$

For $k = 2$ (two hits) this further simplifies to

$$\text{LCP} = 1 - \exp(-q \cdot D) - q \cdot D \exp(-q \cdot D). \qquad 5\text{-}4$$

This equation can be solved for q^* in the same way that the one-hit approximation was solved, substituting equation 5-4 into column E of table 5-7. The maximum-likelihood method yields

$$q^* = 0.0037.$$

Again, knowing that when $x \ll 1$, $\exp(-x) \approx 1 - x$, when $(q \cdot D) \ll 1$, equation 5-4 further simplifies to

$$\text{LCP} = (q \cdot D)^k / k!$$
$$= (q \cdot D)^2 / 2, \qquad \qquad 5\text{-}5$$

and the dose associated with a 10^{-6} excess risk is

$$10^{-6} = (0.0037 \times D)^2 / 2,$$

or

$$10^{-6} = 6.8 \times 10^{-6} \times D^2.$$

Then

$$D^2 = 10^{-6} / 6.8 \times 10^{-6}$$
$$= 0.15,$$
$$D = 0.39 \text{ mg/kg/day.}$$

The assumptions made about the mechanisms of carcinogenesis have a nontrivial effect on the results. There are three orders of magnitude difference between the results of the one-hit and two-hit models for this data set.

Multistage Model. Fitting a two-stage model to the above data set provides an equation to estimate how cancer is caused, assuming that the mechanisms are the same at high and low dose and that two discrete, sequential, and independent stages adequately describe carcinogenesis. As discussed above, these assumptions are problematic. But if they hold, the calculation is as follows, using the "Crump formulation" (Crump 1981), a useful formula for approximating the two-stage model given linear to sublinear data.

The general formula for the multistage approximation is

$$\text{LCP}(D) = 1 - \exp - (\Sigma q_i \cdot D^i). \qquad \qquad 5\text{-}6$$

Since we are looking specifically at the two-stage model, the formula needed is

$$\text{LCP}(D) = 1 - \exp - (q_1 \cdot D^1 + q_2 \cdot D^2). \qquad \text{5-7}$$

In this case, there are two different qs, one for each stage. In order to solve this using the method described above, equation 5-7 is used in column E of table 5-7, and the second q is allowed to change when solving for the maximum likelihood of the first. This yields

$$q_1^* = 0.00042,$$

$$q_2^* = 2.6 \times 10^{-6}.$$

Once the maximum likelihood values of q_1 and q_2 have been calculated, the low-dose approximation is applied, using the observation that for very small values, $q_1 D \gg q_2 D^2$. The formula for low-dose extrapolation, then, is

$$\text{LCP}(D) = 1 - \exp - (q_1 \cdot D),$$

and since $q_1 \cdot D \ll 1$, this equation simplifies to

$$\text{LCP}(D) = q \cdot D. \qquad \text{5-8}$$

Using this to extrapolate to low dose and low probability yields

$$10^{-6} = 0.0021 \times D,$$

so

$$D = 10^{-6}/0.0021$$

$$= 4.8 \times 10^{-4} \text{ mg/kg/day}.$$

Clearly, the different assumptions about the mechanisms of carcinogenesis lead to dramatically different interpretations of the laboratory data. A number of alternative approaches can be used, such as a Weibull model, or curve fitting using a best-fit polynomial. Space constrains us to these three models, but we encourage the reader to explore alternative approaches.

Solution 5-2c

This part of the question actually asks for two quite different solutions at the same time. The first is to compare a lifetime dose, for which we have a number, to a dose that lasts for a fraction of the lifetime. The second is to compare an oral dose (*gavage* means administered directly to the test animal through a tube to the stomach) to an inhaled dose.

Comparing doses at different levels and via different exposure routes requires a dose metric. The standard dose metric used in many risk analyses is that of the area under the curve of dose against time, that is, the integral of dose as a function of time. For example, using this metric, an exposure of 40 mg/day for 10 days is considered to be equivalent to 10 mg/day for 40 days. While this is a useful simplifying assumption, it is subject to the criticism that higher doses may have convex effects: in other words, the 40 mg/day for 10 days might be more toxic than 10 mg/day for 40 days. There are some experiments that seem to confirm this (Cox 1996), but not enough that better dose metrics are consistently available.

Using the area under the curve method, the lifetime equivalent human dose can be compared to the dose represented by the fraction of life spent under exposure conditions as follows (ADD = average daily dose):

$$\text{ADD}_{\text{exposure}} = \text{ADD}_{\text{lifetime}}/f_{\text{exposed}},$$

where

$$f_{\text{exposed}} = \text{fraction of lifetime exposed (unitless, } 0 \leq f \leq 1)$$

$$= \text{YpL} \times \text{WpY} \times \text{HpW}/L.$$

Here, for this example, YpL is years exposed over the lifetime (30 years), WpY is the fraction of weeks exposed (50/52), HpW is hours exposed per week (40/168), and L is the lifetime in years (71 years). (This reflects the case of an individual with stellar attendance at a long-term job.)

We use the following equation to relate the exposure dose to an equivalent lifetime dose, or what the actual dose would be if it were spread across the lifetime:

$$ADD_{exposure} = ADD_{lifetime}/(30 \text{ years}) \times (50/52)$$

$$\times (40/168)/(76 \text{ years})$$

$$= ADD_{lifetime}/0.090.$$

The next step is to compare the concentration of benzene in the air to the daily inhaled dose. One key assumption made here is that the uptake of benzene through the rat stomach is the same as that through human lungs. That is to say an *applied* dose, which is the dose that gets to the barrier (stomach lining, skin, or lungs) results in the same *internal* dose (amount that gets through this barrier) and *delivered* dose (dose available to the organs of concern) regardless of which barrier is relevant. A second assumption is that absorption across a given barrier is the same for the test animal as for humans.

In this case, the air concentration of benzene that applies to $ADD_{exposure}$ is found from

$$C_{b,\,air} = (1/IR) \times BW_{human} \times [ml_b/mg_b] \times ADD_{exposure},$$

where $C_{b,\,air}$ is the concentration of benzene in the air (in ml of benzene per m^3 air), IR is the adult human inhalation rate (20 m^3 per day), and ml_b/mg_b compares benzene weight (C_6H_6 has a molecular weight of 78 g/mole) to volume (0.313 ml/mg). Thus the following equation can be used to fill in the rest of table 5-7:

$$C_{b,\,air} = (1/20 \text{ m}^3/\text{day}) \times 71 \text{ kg} \times (0.313 \text{ ml/mg})$$

$$\times (ADD_{lifetime}/0.90)$$

$$= 12 \times ADD_{lifetime},$$

so

$$ADD_{lifetime} = C_{b,\,air}/[12 \text{ ppm}/(\text{mg/kg/day})].$$

Since $C_{b,\,air}$ in this case is given as 1 ppm,

$$ADD_{lifetime} = 1 \text{ ppm}/[12 \text{ ppm}/(\text{mg/kg/day})]$$

$$= 0.083 \text{ mg/kg/day}.$$

To perform these calculations you will need a number of conversion factors. You can easily derive the following under standard conditions (i.e., room temperature, or 298 K or 25°C, and 1 atmosphere).

$$1 \text{ mg/m}^3 = 24.45/(\text{MW}) \text{ ppm,}$$

where MW is the molecular weight. Similarly,

$$1000 \text{ mg/m}^3 = d \times (\text{ml/m}^3),$$

where d is the density in grams per milliliter.

This means that the thirty-year exposure to 1 ppm is equivalent to a lifetime exposure of 0.083 mg/kg/day. We can now find the numerical answer to the problem by plugging the numbers into the three models.

One-hit (equation 5-3):

$$\text{LCP}(D) = q^* \cdot D$$
$$= 0.0015 \cdot 0.083$$
$$= 1.2 \times 10^{-4}.$$

Two-hit (equation 5 5):

$$\text{LCP}(D) = (q^* \cdot D)^2/2$$
$$= (0.0037 \cdot 0.083)^2/2$$
$$= 4.7 \times 10^{-8}.$$

Two-stage (equation 5-8):

$$\text{LCP}(D) = q^* \cdot D$$
$$= 0.00042 \cdot 0.083$$
$$= 3.5 \times 10^{-5}.$$

Problem 5-D. The Cost of Better Data

Suppose that you want to test the hypothesis that the linear extrapolation for benzene is correct with real data at realistic doses—around 0 to 0.001 mg/kg (human exposure). Calculate the LCP for this dose. How many mice would you need to test if you wanted your results to differ from zero at the 95% confidence level? If tests cost about $10,000 per rat, how much would such an analysis cost?

Problem 5-E. Model-Free Extrapolation

The EPA's proposed extrapolation technique is shown in figure 5-5. It requires generating a curve that represents a 95% upper bound on the response data (the upper dashed curved in the figure). From this the dose associated with a 95% upper bound risk of 0.1 (or LED-10) is calculated. Extrapolation to low dose is linear from the LED-10. Use this method to find the dose associated with a 10^{-6} increased cancer risk for benzene. How does the result differ from the numbers found using the other models? Which numbers do you find more credible or useful? Why?

Problem 5-F. Variation in Cancer Susceptibility

The problem is based on Finkel 1990. Recent research indicates that there are significant differences in carcinogenesis susceptibility, both across and within gender, age, and ethnic groups (Perera 1997). Suppose that for Kil-EZ, a 10^{-6} risk calculated using the two-stage model is believed to represent the geometric mean of individual risks at exposure of 3.5×10^{-4} mg/kg/day. In addition, suppose that because of differences in susceptibility in a population of ten million, risk is distributed lognormally with a geometric standard deviation of 5×10^{-6}. Further suppose that 10^{-5} is considered an "acceptable" level of risk to this population.

   a. Calculate the dose associated with the 10^{-6} risk.
   b. Calculate the expected number of deaths in the population if all individuals were exposed to 3.5×10^{-4} mg/kg/day.
   c. Calculate the number of individuals who, because of higher susceptibility, are above the 10^{-5} "acceptable risk" at exposure of 3.5×10^{-4} mg/kg/day.

d. Calculate the percentage of the expected excess deaths in the population that will occur in this susceptible portion of the population.

e. Discuss the potential policy ramifications of susceptibility. Discuss how the policy ramifications might differ if (i) the susceptible individuals are identifiable versus not identifiable, and (ii) susceptible individuals are spread throughout the population versus tending to be the very young and the elderly.

Problem 5-G. Variation in Sensitivity and Exposure

Repeat problem 5-F using Monte Carlo analysis, with the additional assumption that exposures are distributed lognormally, with a geometric mean of 3.5×10^{-4} mg/kg/day and a geometric standard deviation of 13.2×10^{-4} mg/kg/day.

Problem 5-H. Additional Data Set

Repeat problem 5-2 with the 1,3-butadiene data set in table 5-2 (problem 5-C).

Problem 5-I. Exact Two-Stage Formulation

This problem was contributed by Weihsueh Chiu. Cox (1995) argued that the traditional simplification of a two-stage model misidentified the stages. He proposed an alternative formulation, the "exact two-stage" (ETS) formulation of the model. Chiu et al. (1999) have shown that, under a range of conditions, the ETS model fits data sets at least as well as, and sometimes substantially better than, the traditional formulation. This problem explores the ETS formulation, and compares it to the traditional version.

The tumor probability for a dose x is given by the traditional two-stage model by $P(x)[\text{trad}] = 1 - \exp(-(q_0 + q_1 x + q_2 x^2))$.

The "exact two-stage" model gives a tumor probability for dose x of

$$P(x)[\text{exact}] = 1 - \exp(-(a_1 + b_1 x)$$
$$\times (1 - (1 - \exp(-(a_2 + b_2 x)))/(a_2 + b_2 x))).$$

Here, tumor probability, $P(x)$ simply means the number of animals with tumors, given a dose x divided by the total number of animals that are given dose x. The variables q_0, q_1, and q_2 are the parameters of the traditional model, and a_1, b_1, a_2, and b_2 are the parameters of the exact model.

For each dose level x_i, we assume we have Y_i responders out of N_i animals. The probability of this happening is given by the binomial distribution as

$$\text{PROB}_i(n_i C y_i) \times P(x_i)^{Y_i} \times (1 - P(x_i))^{(N_i - Y_i)},$$

where the first term is the binomial function $N_i!/((N_i - Y_i)! \times Y_i!)$ and ! is the factorial function.

The "likelihood" of a model for m doses is the product

$$\text{likelihood} = \text{PROB}_1 \times \text{PROB}_2 \times \cdots \times \text{PROB}_m.$$

Maximizing the likelihood by changing the parameters gives the maximum-likelihood estimate (MLE) parameters of the model.

a. One of the standard methods for fitting parameters of a model is the use of the maximum likelihood. This means the best-fitting parameters are the set that maximizes the likelihood. For the vinyl chloride data given in table 5-8, find the maximum-likelihood parameters of the traditional model as well as of the exact model. Save the value of this maximum likelihood.

b. Using the low-dose extrapolation from this model, find the dose that corresponds to a 10^{-5} probability in each model. (We used the "GOAL SEEK" function in Excel.)

c. Now we need to understand the uncertainty in this model fit. One way to do so is to use a Monte Carlo simulation.

   i. Generate 10 random data sets using the probabilities in the data (i.e., use Y_i/N_i as the binomial probability, and N_i as the number of samples).

Table 5-8 Information on Animal Carcinogenesis Studies for Three Chemicals

Name	Abbreviation	Pathway	Response	Reference
Vinyl chloride	VCL	Inhalation	Liver angiosarcoma	FoodSafety
Benzene	BENZ	Oral gavage	Squamous cell carcinomas	Crump and Allen
Ethylene thiourea	ETU1	Ingestion	Fetal anomalies	FoodSafety

Abbreviation	Dose response (no. of responders / total)						Dose unit
VCL	0	50	250	500	2500	6000	ppm
	0/58	1/59	4/59	7/59	13/59	13/60	
BENZ	0	25	50	100	—	—	mg/kg/d
	0/50	4/50	20/50	37/50	—	—	
ETU1	0	5	10	20	40	80	mg/kg/d
	0/167	0/132	1/138	14/81	142/178	24/24	

ii. Repeat (a) and (b) for each of the 10 data sets. What is the range of doses you get for 10^5 risk? Do the models give the same results?

d. In order to assess how well the model fits the data, we use another Monte Carlo simulation. This time, however, you use the maximum-likelihood fit probabilities as your Monte Carlo probability parameters:

i. Generate 10 random data sets using the fit probabilities from (a). That is, use $P(x_i)$[trad] or $P(x_i)$[exact] as your binomial probability. You should still use N_i as your number of samples.

ii. Repeat (a) for each data set.

iii. Sort the ln(likelihoods) you get for your Monte Carlo data sets in increasing order. Where does the maximum likelihood from fitting to the data rank in this Monte Carlo set? If it is one of the lower likelihoods, then the fit is a "bad fit." In particular, if it is the lowest or second lowest, than the models is "ruled out at about 10% confidence." If it is one of the higher likelihoods, then it is a "good fit." How does the goodness of fit for the different models compare?

e. Repeat for the other two data sets in table 5-8. How do the models compare?

Problem 5-3. Noncarcinogenic Effects: The EPA Approach

For a certain pesticide showing no carcinogenic effects, the following data were found with a reproductive toxicity study in rats (dose levels of 0, 200, 500, or 700 ppm). (Only one study has been completed for noncarcinogenic effects.) A reproductive toxicity no observable effect level (NOEL) of 200 ppm was found (14 mg/kg/day in males and 16 mg/kg/day in females). At 500 ppm (33 and 44 mg/kg/day, males and females, respectively), the following effects were observed: decreased maternal body weight and/or body weight gain during gestation in both parent and offspring generations; reduced premating body weight gains in the offspring generation; increased duration of gestation in both offspring and third-generation dams (rat mothers); reduced prenatal viability (fetal and litter); reduced litter size, increased number of stillborn pups; reduced pup and litter postnatal survival; and decreased pup body weights throughout lactation. Male fertility was reduced in the offspring generation at concentration doses of 500 and 700 ppm, with degeneration and/or atrophy of the testes and other reproductive tissues. In addition, developmental toxicity was observed in the rats at 25 mg/kg/day, evidenced as decreased fetal weights and increased variations. The NOEL for developmental toxicity in the rat was 10 mg/kg/day. The EPA Reference Dose (RfD) Committee has recommended that an RfD be established based on these data.

Solution 5-3

The EPA's RfDs are conceptually similar to the Occupational Safety and Health Administration's "permissible exposure limits" (PELs) and a number of other regulatory estimates of "safe doses." These are doses below which the agencies feel that lifetime exposure is acceptable, and are based on probability of excess death as the result of exposure. It is important to note that, while the methods used for calculating these numbers are similar, the decisions about what is "safe" can vary as a function of agency and legislation. For example, OSHA's level of concern is 1:1000, while EPA often uses 1:100,000 or 1:1,000,000 (OSHA 1989; U.S. EPA 1980).

The animal experiments for noncarcinogens are similar to those for carcinogenesis studies, but risk assessment for noncarcinogenic chronic risks differs from that for carcinogens in a number of ways. Perhaps most important is the assumption that there is a threshold level that the body can tolerate. The steps leading to cancer are generally assumed to be irreversible unless there is case-specific information indicating otherwise. In contrast, the body appears to be able to deal with some amount of noncarcinogenic chemicals (up to a threshold), so rather than calculate doses associated with acceptable risk levels, acceptable doses at which no effects are anticipated are often calculated.

To find acceptable doses for noncarcinogenic effects, one of four experimental dose levels is used as a reference, the no observable effect level (NOEL), the no observable adverse effect level (NOAEL), the lowest observable effect level (LOEL), or the lowest observable adverse effect level (LOAEL). To deal with uncertainty, a variety of "safety factors" are used to arrive at conservative estimates. These typically include the following:

- A factor of 1 to 10 to deal with uncertainty in extrapolating between species (1 if humans are the subjects). This is applicable in this problem because rat data are being used to estimate human doses.
- A factor of 1 to 10 for variation within species. This accounts for heterogeneity of the population, and differences between men and women. The default is often 1, unless there is evidence of variability within the population.
- A factor of 10 when only a LOEL or a LOAEL is available, and not a NOEL or NOAEL. In this case, a NOEL is available, so no adjustment is needed.
- A factor of 1 to 10 for multigenerational effects.
- A factor of 1 to 10 to account for the quality of the data (called the "modification factor").

This example is based on a real EPA finding, where the following conclusions were drawn. There is a NOEL for rats, and there does appear to be an intergenerational effect, but there are no data suggesting intraspecies variability. Consequently, a factor of 10 was used for interspecies comparison, a factor of 3 for multigenerational effects was indicated, and a factor of 10 was deemed reasonable to account for this being the only available study. Thus the reference dose would be

$$RfD = NOEL/(uncertainty\ factors \times modification\ factor)$$
$$= 200\ ppm/(10 \times 3 \times 10)$$
$$= 0.7\ ppm.$$

Problem 5-J. Additional Noncancer End Points

Suggest a reference dose given the following data: a three-generation reproduction study (dose levels of 0, 1000, or 10,000 ppm) in which the LOEL for offspring systemic toxicity is 10,000 ppm (661 mg/kg/day) based on reduced pup body weights late in lactation and a reproductive toxicity study with a NOEL of 63 mg/kg/day (1000 ppm). The NOEL for developmental effects in one species (rat) was greater than 1000 mg/kg/day. The NOEL for developmental toxicity in the rabbit was 40 mg/kg/day. The LOEL in the rabbit developmental toxicity study was 200 mg/kg/day based on increased resorptions (viable ova that are reabsorbed by the mother). Because the NOELs for the rabbit developmental toxicity and reproduction studies were similar, the agency used the NOEL from the reproduction study. The NOEL from the reproduction study was used to set the reference dose because the route of exposure was dietary, which was deemed more appropriate for a chronic dietary risk assessment than short-term gavage dosing (which is the method of exposure in the developmental toxicity study).

Problem 5-K. Formaldehyde

In 1981, final results were obtained from the experimental exposure of rats to formaldehyde in air. The results are shown in table 5-9.

a. How many cancers would have been expected at the lower exposure levels if the incidence of cancer at the highest exposure level had been extrapolated downward linearly?

b. Measured formaldehyde levels in U.S. mobile homes and houses insulated with urea foam range from 0.02 to 4.2 ppm. Let us

Table 5-9 Formaldehyde Test Data

Exposure (ppm)	Incidence of nasal carcinomas (cancer cases / exposed no. of rats)
2.0	0/236
5.6	2/235
14.3	103/232

assume an average exposure of 0.50 ppm. Assuming that humans are just as susceptible as rats and that ten million people are exposed, how many nasal cancers would be predicted by linear extrapolation from the highest rat exposure levels over the lifetimes of these people?

c. What would be the prediction of total human cancers for a linear extrapolation from the 5.6 ppm rat-exposure level?

References

Aaronson, S. A. (1993). "Growth factors and cancer," Science 254:1146–53.

Ames, B. N. and Gold, L. S. (1991). "Carcinogenesis mechanisms—the debate continues." *Science* 252 (5008):902.

Anderson, E., Deisler, P., McCallum, D., St. Hilaire, C., Spitzer, H., Strauss, H., Wilson, J., and Zimmerman, R. (1993). "Key issues in carcinogenic risk assessment guidelines, Society for Risk Analysis." *Risk Analysis* 13(4):379–82.

Chiu, W. A., Hassenzahl, D. M., and Kammen, D. M. (1999). "A comparison of regulatory implications of traditional and exact two-stage dose-response models." *Risk Analysis* 19(1):37–44.

Cox, L. A, (1995). "Reassessing benzene risks using internal doses and Monte-Carlo uncertainty analysis." *Environmental Health Perspectives* 104(Suppl. 6):1413–29.

Croy, R. G. (1993). "Role of chemically induced cell proliferation in carcinogenesis and its use in health risk assessment." *Environmental Health Perspectives* 101 (Supplement 5):289–302.

Crump, K. S. (1981). "An improved procedure for low-dose carcinogenic risk assessment from animal data." *J. Environmental Pathology and Toxicology* 5.

Crump, K. S. and Allen, B. C. (1984). *Quantitative Estimates of the Risk of Leukemia from Occupational Exposures to Benzene. Final Report to the Occupational Safety and Health Administration.* Ruston, LA: Science Research Systems.

Finkel, A. M. (1990). "A simple formula for calculating the "mass density" of a lognormally distributed characteristic: applications to risk analysis." *Risk Analysis* 10(2):291–301.

Gold, L. S., Manley, N. B., and Ames, B. N. (1992). "Extrapolation of carcinogenicity between species: Qualitative and quantitative factors." *Risk Analysis* 12(4):579–88.

Gots, R. E. (1993). *Toxic Risks: Science, Regulation and Perception.* Ann Arbor, MI: Lewis Publishers.

Infante, P. F. (1993). "Use of rodent carcinogenicity test results for determining potential cancer risk to humans." *Environmental Health Perspectives* 101(Suppl 5):143–48.

Knudson, A. G., Jr, and Strong, L. C. (1972). "Mutation and cancer: neuroblastoma and pheochromocytoma." *American Journal of Human Genetics* 24(5):514–32.

Krewsy, D., Gaylor, D. W., Soms, A. P., and Szyszkowicz, M. (1993). "An overview of the report: Correlation between carcinogenic potency and the maximum tolerated dose: Implications for risk assessment." *Risk Analysis* 13(4):383–98.

National Research Council. (1993). *Issues in Risk Assessment: I, Use of the Maximum Tolerated Dose in Animal Bioassays for Carcinogenesis.* Washington DC: National Academy Press.

Perera, F. P. (1997). "Environment and cancer: who are susceptible?" *Science* 278(November 7):1068–73.

U.S. Department of Energy. (1990). *Indoor Radon and Decay Products: Concentrations, Causes, and Control Strategies.* DOE/ER-0480P. Washington DC.

U.S. EPA. (1997). "Proposed guidelines for carcinogen risk assessment." *Federal Register* 61(79) (April 23) 17960–18011.

U.S. EPA. (1980). "Guidelines and methodology used in derivation of the health effect assessment chapter of the consent decree water criteria document." *Code of Federal Register* 45:79347–49.

U.S. OSHA. (1989). "Air contaminants, final rule." 54:2332–983.

Watanabe, K., Bois, F., and Zeise, L. (1992). "Interspecies extrapolation: A reexamination of acute toxicity data." *Risk Analysis* 12(2):301–10.

Weinberg, R. A. (1993). "Tumor suppressor genes." *Science* 254:1138–46.

Wilson, J. D. and Anderson, J. W. (1997). "What the science says: How we use it and abuse it to make health and environmental policy." *Resources* 128:5–8.

Wilson, R. and Crouch, E. A. (1987). "Risk assessment and comparisons: an introduction." *Science* 236(4799):267–70.

6

Epidemiology

Introduction

Epidemiological studies are the basis for innumerable public policies. Because they often contain data on populations actually exposed to suspected risk factors, these studies can provide valuable information in many cases. At other times, however, the minute exposure differences between populations can make policy decisions problematic, as effects may be too small to allow clear statistical conclusions. The following problems look at some clear, and some less clear, epidemiological studies, and demonstrate some strengths and limitations of the method.

Problem 6-1. Cigarette Smoking and Cancer

Consider the following (approximate) figures relating to cigarette smoking and cancer: The death rate in the U.S. population of 270 million is 10 deaths per 1000 persons per year. One-fifth of these deaths are from cancer, and one-fourth of the cancer deaths are from lung cancer. Ninety percent of the lung cancer deaths occur among the sixty million Americans who are smokers. Figure 6-1 shows all cancer deaths in the United States over the last fifty years, by type of cancer.

   a. How many deaths are there per year in the United States from all causes? How many cancer deaths? How many lung cancer deaths?
   b. What is the lung cancer death rate among smokers expressed in annual deaths per 1000 smokers? What is the lung cancer death rate among nonsmokers expressed in annual deaths per 1000 nonsmokers? Compute the ratio of these rates, smokers' to nonsmokers'.
   c. Of the lung cancer deaths occurring among nonsmokers, at least one fifth are thought to be attributable to environmental ("second-hand") tobacco smoke. How many deaths per year would

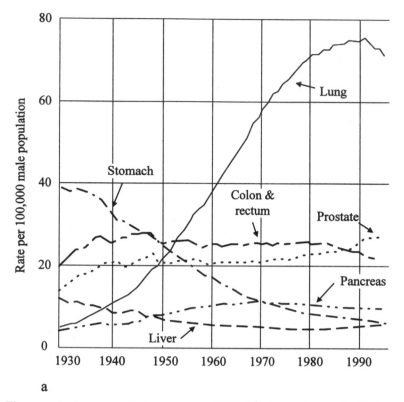

Figure 6-1. Cancers, all types, since 1930 (a) for males and (b) for females. After American Cancer Society 1998.

this be? Given that the average smoker consumes 15 cigarettes per day, and that there are 60 million smokers in the United States, how many cigarettes are associated with each smoking-induced lung cancer death? How many with each second-hand smoker death?

d. Benzo[a]pyrene (B(a)P) is an organic compound that is believed to be a human carcinogen. It is generated when coal, wood, and other plant materials (including tobacco) are burned (Harte et al. 1991). The EPA "unit risk factor" (URF) is used to calculate the individual lifetime cancer risk (LCR) from inhaling a chemical using the formula

$$LCR = (\text{lifetime average } \mu g/m^3) \times URF.$$

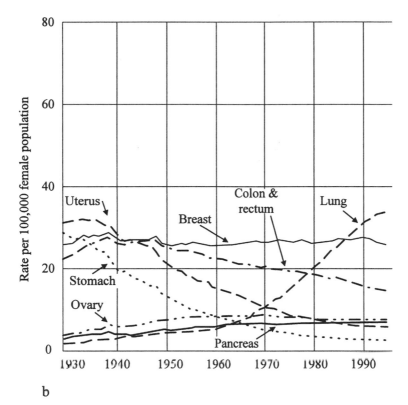

b

Figure 6-1. Continued.

    i. OSHA's advisory level for B[a]P is 0.2 $\mu g/m^3$. This means that individuals should not be exposed to B[a]P concentrations above this level. Does exposure to B[a]P through cigarette smoke exceed this? Evaluate this in three ways, first assuming that the standard is for average daily exposure, second assuming that it is an average over eight hours, and finally assuming that it is a threshold that should never be reached.

    ii. Use the information in table 6-1 to estimate how much of the lung cancer impact of tobacco smoke may plausibly be attributed to its benzo[a]pyrene content.

   iii. Discuss the uncertainties in your calculation for (ii).

e. We originally considered writing part (c) as "Given that the 60 million U.S. smokers consume an average of 15 cigarettes per day,

Table 6-1 Information Needed for Problem 6-1

Unit risk factor for B[a]P	$1.75 \times 10^{-2}/\mu g/m^3$
B[a]P inhaled, per cigarette	$0.03\ \mu g$
Average breath volume	1.2 liters
Average breathing rate	10–15 breaths per minute

how many cigarettes does it take to kill a smoker?" Why is this wording problematic?

Solution 6-1

Solution 6-1a

Deaths from all causes:

$$(270 \text{ million people}) \left(\frac{10 \text{ deaths}}{1000 \text{ persons} \cdot \text{year}} \right)$$

$$= 2.7 \text{ million deaths/year.}$$

Deaths from cancer:

$$(2.7 \text{ million deaths/year}) \left(\frac{1 \text{ death from cancer}}{5 \text{ deaths}} \right)$$

$$= 540,000 \text{ cancer deaths/year.}$$

Deaths from lung cancer:

$$(540,000 \text{ cancer deaths/year}) \left(\frac{1 \text{ lung cancer death}}{4 \text{ cancer deaths}} \right)$$

$$= 140,000 \text{ lung cancer deaths/year.}$$

Solution 6-1b

Ninety percent of the lung cancer deaths occur among smokers:

$$(140,000 \text{ lung cancer deaths/year})(0.90) = 130,000 \text{ deaths/year.}$$

So

$$\text{death rate} = \left(\frac{130{,}000 \text{ deaths/year}}{60 \text{ million smokers}} \right)$$

$$= 2.2 \text{ deaths}/1000 \text{ person-years.}$$

Ten percent of the lung cancer deaths occur among nonsmokers:

$$(140{,}000 \text{ lung cancer deaths/year})(0.10) = 14{,}000 \text{ deaths/year.}$$

The total number of nonsmokers is 270 million − 60 million = 210 million. So

$$\text{death rate} = \left(\frac{14{,}000 \text{ deaths/year}}{210 \text{ million nonsmokers}} \right)$$

$$= 0.067 \text{ deaths}/1000 \text{ person-years.}$$

Solution 6-1c

Deaths from second-hand smoke are given by

$$\left(\frac{1}{5} \right)(14{,}000) = 2800 \text{ deaths/year.}$$

The number of cigarettes associated with each smoking-induced lung cancer death is

$$(60 \text{ million smokers})\left(\frac{15 \text{ cigarettes}}{\text{smoker} \cdot \text{day}} \right)\left(\frac{365 \text{ days}}{\text{year}} \right)$$

$$= 3.3 \times 10^{11} \text{ cigarettes/year,}$$

so for smokers,

$$\frac{3.3 \times 10^{11} \text{ cigarettes}}{130{,}000 \text{ smoker lung cancer deaths/year}}$$

$$= 2.5 \text{ million cigarettes per smoker–lung cancer death,}$$

and for nonsmokers,

$$\frac{3.3 \times 10^{11} \text{ cigarettes/year}}{2800 \text{ second-hand smoker lung cancer deaths/year}}$$

$$= 120 \text{ million cigarettes per nonsmoker–lung cancer death}$$
from second-hand smoke.

Solution 6-1d(i)

The URF for average daily exposure is calculated by dividing total daily exposure by total inhalation over the day. Total daily exposure is

$$\left(\frac{0.03 \ \mu\text{gB[a]P}}{\text{cigarette}} \right) \left(\frac{15 \text{ cigarettes}}{\text{day}} \right) = 0.45 \ \mu\text{gB[a]P per day.}$$

Total daily inhalation is

$$\left(\frac{12.5 \text{ breaths}}{\text{minute}} \right) \left(\frac{24 \times 60 \text{ minutes}}{\text{day}} \right) \left(\frac{1.2 \text{ liters}}{\text{breath}} \right)$$
$$= 22,000 \text{ liters per day.}$$

So average daily exposure is

$$\left[\left(\frac{0.45 \ \mu\text{gB[a]P}}{\text{day}} \right) \Big/ \left(\frac{22,000 \text{ liters}}{\text{day}} \right) \right] \left[\frac{1000 \text{ liters}}{\text{m}^3} \right]$$
$$= 2.2 \times 10^{-2} \ \mu\text{gB[a]P/m}^3,$$

which is an order of magnitude below the OSHA advisory level.

If the OSHA standard is an average over eight hours the calculation is as follows. Assume that the individual smokes the fifteen cigarettes over fifteen waking hours, or one cigarette per hour (note that average hourly exposure with this assumption will be equal to eight-hour average exposure). Hourly exposure is

$$\left(\frac{0.03 \ \mu\text{gB[a]P}}{\text{cigarette}} \right) \left(\frac{1 \text{ cigarettes}}{\text{hour}} \right) = 0.03 \ \mu\text{gB[a]P per hour.}$$

Hourly inhalation is

$$\left(\frac{12.5 \text{ breaths}}{\text{minute}}\right)\left(\frac{60 \text{ minutes}}{\text{hour}}\right)\left(\frac{1.2 \text{ liters}}{\text{breath}}\right) = 900 \text{ liters per hour.}$$

So average hourly exposure is

$$\left[\left(\frac{0.03 \text{ } \mu\text{gB[a]P}}{\text{hour}}\right)\middle/\left(\frac{900 \text{ liters}}{\text{hour}}\right)\right]\left[\frac{1000 \text{ liters}}{\text{m}^3}\right]$$

$$= 3.3 \times 10^{-2} \text{ } \mu\text{gB[a]P/m}^3,$$

still about an order of magnitude below the limit.

In the final case, where the OSHA standard is a threshold that should never be reached, the only relevant exposure is during an actual puff:

$$\left(\frac{0.03 \text{ } \mu\text{gB[} \alpha \text{]P}}{\text{cigarette}}\right)\left(\frac{\text{cigarette}}{10 \text{ puffs}}\right)\left(\frac{\text{puff}}{1.2 \text{ liters}}\right)\left(\frac{1000 \text{ liters}}{\text{m}^3}\right)$$

$$= 2.5 \text{ } \mu\text{g/m}^3.$$

This is two orders of magnitude above the allowable workplace concentration.

Solution 6-1d(ii)

First, estimate the total lifetime intake of B[a]P from cigarette smoke. Assume that a smoker starts, on average, at fifteen years of age, and smokes for the remainder of a sixty-year lifetime:

$$\left(\frac{15 \text{ cigarettes}}{\text{day}}\right)\left(\frac{365 \text{ days}}{\text{year}}\right)\left(\frac{45 \text{ years of smoking}}{\text{smoker-lifetime}}\right)$$

$$\left(\frac{0.03 \text{ } \mu\text{g B[a]P}}{\text{cigarette}}\right)$$

$$= 7400 \text{ } \mu\text{g B[a]P/smoker-lifetime.}$$

Next, estimate the total amount of air that the smoker inhales over his lifetime:

$$\left(\frac{1.2 \text{ liters}}{\text{breath}}\right)\left(\frac{12.5 \text{ breaths}}{\text{minute}}\right)\left(\frac{24 \times 60 \text{ minutes}}{\text{day}}\right)$$

$$\times \left(\frac{365 \text{ days}}{\text{year}}\right)\left(\frac{60 \text{ years}}{\text{smoker-lifetime}}\right)$$

$$= 4.7 \times 10^8 \text{ liters of air inhaled per smoker-lifetime.}$$

Putting these together, estimate the lifetime average exposure to B[a]P from cigarette smoke:

$$\left[\left(\frac{7400 \ \mu g}{\text{lifetime}}\right)\bigg/\left(\frac{4.7 \times 10^8 \text{ liters}}{\text{lifetime}}\right)\right]\left(\frac{1000 \text{ liters}}{\text{m}^3}\right)$$

$$= 1.6 \times 10^{-2} \ \mu g/\text{m}^3.$$

Then use the URF to calculate the LCR:

$$\text{LCR} = 1.6 \ \mu g/\text{m}^3 \times 1.75 \times 10^{-2} \ (\mu g/\text{m}^3)^{-1} = 0.028;$$

or a 2.8% lifetime cancer risk associated with B[a]P in cigarette smoke.

Problem 6-A. The Heavy Smoker

Recalculate 6-1c and 6-1d for a heavy smoker. Explain any assumptions you need to make.

Problem 6-B. All Deaths in the United States

Estimate the expected number of deaths from heart disease and cancer this year in a class of twenty college seniors. Also, what is the class size for 1 expected cancer fatality this year? Why may these estimates be inappropriate? Calculate upper and lower bounds for your estimates.

Problem 6-2. Risk in a Time of Cholera

Today, we understand that cholera is caused by a water-borne bacterium that is common in water contaminated by human and animal wastes. However, before the 1900s, the cause was not known, and cholera epidemics were a serious public health threat. In 1854, a physician named John Snow intuited an association between drinking water supply and cholera in London. He knew that of two major water suppliers who got water from the Thames River, one (Lambeth) pumped from upstream of London, while the other (Southwark and Vauxhall) pumped from downstream. Snow found data (summarized in table 6-2) on cholera deaths in houses hooked up to each of the water sources, as well as total cholera deaths and total houses in London.

a. Test the following hypotheses:

   i. That the risk from drinking water from Southwark and Vauxhall is the same as that from Lambeth
   ii. That the risk associated with drinking water from Southwark and Vauxhall is the same as that for the rest of London
   iii. That the risk associated with drinking water from Lambeth is the same as that for the rest of London

b. Calculate the relative risk of getting water from Lambeth and from Southwark and Vauxhall.

c. Snow was able to find complete information from company records on which houses got water from each company. Because both companies had pipes throughout London, next-door neighbors often got their water from different sources. Snow found no indicators (such as income, social status, location) that predicted whether a particular house got its water from one company or the

Table 6-2 London Cholera Data, 1853–54

Water company	Number of houses	Cholera deaths
Southwark and Vauxhall	40,046	1263
Lambeth	26,107	98
Rest of London	256,423	1422

other. Why is this an important observation? If you did not know this, how would it impact your evaluation of the data in table 6-2?

d. Assume that you find evidence that the undertakers misdiagnosed cholera as the cause of death in 50% of cases. How would this affect Snow's findings?

Solution 6-2

Solution 6-2a

The tools to test these hypotheses should be quite familiar by now. Since the sample size is large, the Z-score is appropriate. From equation 3-7c,

$$Z = \frac{\bar{x} - np}{\sqrt{np(1 - p)}}.$$

In case (i), $\bar{x} = 1263$, the observed number of deaths for the Southwark and Vauxhall group, $n = 40{,}046$, the number of people in the Southwark and Vauxhall group, and $p = 98/26{,}107 = 0.0038$, the hypothesized probability of death if Southwark and Vauxhall is the same as the rest of London. Substituting,

$$Z = \frac{\bar{x} - np}{\sqrt{np(1 - p)}} = \frac{1263 - 40{,}046 \cdot 0.0038}{\sqrt{40{,}046 \cdot 0.0038(1 - 0.0038)}} = 90.$$

If we want to test for a 99% confidence level, the Z-table in appendix A shows that we can reject the hypothesis that the cholera death rate among Southwark and Vauxhall customers is the same as that for Lambeth customers.

For case (ii), $\bar{x} = 1263$, the observed number of deaths for the Southwark and Vauxhall group, $n = 40{,}046$, the number of people in the Southwark and Vauxhall group, and $p = 1422/256{,}423 = 0.00555$, the hypothesized probability of death if Southwark and Vauxhall is the same as the rest of London. Plugging in the numbers,

$$Z = \frac{\bar{x} - np}{\sqrt{np(1 - p)}} = \frac{1263 - 40{,}046 \cdot 0.00555}{\sqrt{40{,}046 \cdot 0.00555(1 - 0.00555)}} = 70.$$

As above, we can reject the hypothesis that the cholera death rate among Southwark and Vauxhall customers is the same as that for the rest of London.

Finally, for case (iii),

$$Z = \frac{\bar{x} - np}{\sqrt{np(1 - p)}} = \frac{98 - 26{,}107 \cdot 0.00555}{\sqrt{40{,}046 \cdot 0.00555(1 - 0.00555)}} = -3.9.$$

Even in this case, we can reject the hypothesis.

Solution 6-2b

The relative risk (RR) is the ratio of the risk faced by an exposed group to that for everyone else. In this case, for Southwark and Vauxhall customers,

$$RR = \frac{1263/40{,}046}{1422/256{,}423} = 5.7,$$

and for Lambeth customers,

$$RR = \frac{98/26{,}107}{1422/256{,}423} = 0.68.$$

In words, this means that being a Southwark and Vauxhall customer entailed almost six times the cholera risk of the rest of London, but that being a Lambeth customer meant only about two-thirds the risk. The calculations in solution 6-2a suggest that we can probably believe that the observed differences are real.

Solution 6-2c

Parts (*a*) and (*b*) provide particularly compelling evidence about an *association* between water company hookups and cholera risk, but without this additional qualification, we have no evidence that the water supply company was the *cause* of cholera. For example, what if Southwark and Vauxhall customers were primarily the poor, who had less food and heat, and consequently were more likely to die from cholera? Or what if Lambeth customers lived far away from Southwark and Vauxhall customers, and therefore were isolated from a particular cholera outbreak? Only when we are certain that all other factors are the same, or at least not significant compared to the water supply variable, can we be comfortable with conclusions about water supply as a source of cholera.

Solution 6-2d

Clearly, this attenuates the findings. This is an example of a fundamental limitation of statistical analysis: measurement error will always undermine confidence. Approaches to this problem include the following:

1. Think about the outside case, that deaths in "the rest of London" were underestimated by half, and those for Southwark and Vauxhall were overestimated by half. If the findings still stand, it is a strong indication that there was a real effect.
2. Approach 1 is computationally easy but is a "brute force" approach in the sense that it overestimates the potential error. A more subtle option is to assume some sort of distribution around each of the means and calculate whether there is a significant overlap in the tails.

Problem 6-C. Pooled Data

Pool the three groups into an "all of London" group, and test the hypothesis that "all of London" is the same as "the rest of London," which excludes Southwark and Vauxhall and Lambeth customers. Repeat problem 6-2b using the pooled data. Does it make a difference? Discuss.

Problem 6-D. What Might Be Missing?

List some other reasons that condition 6-2c might be necessary. Suppose that, on average, income among Southwark and Vauxhall customers was 95% of that for Lambeth customers (90% statistical confidence). Would this change your opinion about the validity of your finding? What if income for Southwark and Vauxhall customers was, on average, half of that for Lambeth customers (99% confidence)?

Problem 6-E. Measurement Error

Compute the measurement error effect using approaches 1 and 2 presented qualitatively in solution 6-2d. Suggest and compute an alternative approach. Discuss your findings.

Problem 6-3. Benzene Revisited: The Pliofilm Cohort Study

A considerable amount of epidemiological data on benzene risk is available from a cohort study of a large number of workers at Pliofilm rubber manufacturing facilities from the 1940s through the 1970s (studies continue to follow members of this cohort) (Crump 1996). In addition, the data are consistently refined and reanalyzed as new interpretations of exposure and health effects come in. Lymphatic and hematopoietic cancer (which includes leukemia) are suspected to be caused by benzene.

Table 6-3 contains data on mean cumulative exposure (in ppm-years) for the individuals who were exposed to benzene over the years, person-years of exposure, cases of all lymphatic and hematopoietic cancer deaths, and observed cases of leukemia, as well as the expected death rate from those diseases based on individuals of similar age, background, and so on who were not exposed to benzene.

Table 6-3 Aggregated Pliofilm Cohort Data

Cumulative exposure (ppm-years) mean	Persons years	Total lymphatic and hematopoietic		Leukemia		Total excluding leukemia	
		Observed deaths	Expected deaths per person-year	Observed deaths	Expected deaths per person-year	Observed deaths	Expected deaths per person-year
132	52584	21	2.3×10^{-4}	14	9.0×10^{-5}	7	1.40×10^{-4}

a. Why are the expected death rates different for different exposure levels? Why isn't the rate for these diseases for the overall population used?
b. Calculate the RR for leukemia, for all lymphatic and hematopoietic cancer deaths, and for nonleukemia lymphatic and hematopoietic cancer deaths. Calculate a p score associated with each RR. Why did the researchers consider leukemia separately?

Solution 6-3

Solution 6-3a

The expected death rates in this study were adjusted to match the individuals who were in the study: mostly working-age males. A considerable amount of effort was expended to ensure that the comparison (control) group was similar in age, lifestyle, prior health factors, and so on. An estimate from the entire population would have brought in a number of possible biases, such as differences in susceptibility in the (nonworking) young and elderly, and increasing likelihood of cancer from all causes as a function of age.

Solution 6-3b

We can calculate the expected number of deaths from the expected death rate and the number of individuals exposed:

expected total deaths

$$= 52584 \text{ person-years} \times (2.30 \text{ deaths}/10^4 \text{ person-years})$$
$$= 12.1 \text{ deaths.}$$

From this, a relative risk can be calculated:

$$RR = \text{observed number of deaths}/\text{expected number of deaths}$$
$$= 21/12.1 = 1.74.$$

As in problem 6-2, a Z-score provides an indication of the significance of the relative risk:

$$Z = \frac{21 - 12.1}{\sqrt{52584 \cdot 2.30 \times 10^{-4} \cdot (1 - 2.30 \times 10^{-4})}} = 2.59.$$

This Z-score is associated with $p < 0.005$, suggesting that benzene exposure is associated with increases in lymphatic and hematopoietic cancers.

Repeating the above steps for the other two categories, we can create table 6-4. This table suggests that, while there appears to be a statistically significant increase in lymphatic and hematopoietic cancer deaths, it is entirely attributable to the subset of leukemia deaths.

Problem 6-F. Additional Data

Table 6-5 contains the data from table 6-3 broken down into different exposure categories. For each exposure level, calculate the RR for leukemia, for all lymphatic and hematopoietic cancer deaths, and for non-leukemia lymphatic and hematopoietic cancer deaths. Are any of the findings statistically significant? What do you infer from this?

Problem 6-G. One-Hit Model and Epidemiological Data

Fit a one-hit model of carcinogenesis to the data in table 6-3. Compare your results to those from problem 5-2. Discuss the following issues:

a. Is one of these data sets better than the other? What are the strengths and weaknesses of each?

b. How would you treat these studies if you had to make a decision about regulating workplace benzene exposure? Would you pick one study over the other? Would you try to use both?

Table 6-4 Statistical Test of Aggregated Pliofilm Cohort Data

Category	Total lymphatic and hematopoietic	Leukemia	Total excluding leukemia
Expected deaths	12.1	4.75	7.36
RR	1.74	2.9	0.095
Z-score	2.59	4.25	0.13
$p <$	0.005	0.0001	0.5

Table 6-5 Pliofilm Cohort Data: Total Lymphatic and Hematopoietic, Leukemia, and Total Excluding Leukemia

Cumulative exposure (ppm-years)		Person-years	Total lymphatic and hematopoietic		Leukemia		Total excluding leukemia	
Range	Mean		Observed deaths	Expected deaths per person-year	Observed deaths	Expected deaths per person-year	Observed deaths	Expected deaths per person-year
0–45	11	30482	6	2.02×10^{-4}	3	7.91×10^{-5}	3	1.23×10^{-4}
45–400	151	16320	6	2.35×10^{-4}	4	9.19×10^{-5}	2	1.43×10^{-4}
400–1000	602	4667	3	3.39×10^{-4}	2	1.34×10^{-4}	1	2.03×10^{-4}
> 1000	1341	915	6	4.81×10^{-4}	5	1.97×10^{-4}	1	2.95×10^{-4}
Total	132	52584	21	2.30×10^{-4}	14	9.03×10^{-5}	7	1.40×10^{-4}

Table 6-6 Pliofilm Cohort Data: AMML and Leukemia Excluding AMML

Cumulative exposure (ppm-years)			AMML		Leukemia excluding AMML	
				Expected deaths		Expected deaths
Range	Mean	Person-years	Observed deaths	per person-year	Observed deaths	per person-year
0–45	11	30482	0–2	2.69×10^{-5}	1–3	5.22×10^{-5}
45–400	151	16320	1	3.12×10^{-5}	3	6.07×10^{-5}
400–1000	602	4667	2	4.52×10^{-5}	0	8.84×10^{-5}
> 1000	1341	915	5	6.56×10^{-5}	0	1.31×10^{-4}
Total	132	52584	8–10	3.06×10^{-5}	4–6	5.97×10^{-5}

Problem 6-H. Additional Data

Table 6-6 contains data on acute mylocytic or acute monocytic leukemia (AMML), a subset of leukemias. Repeat problem 6-3(*b*) using these new data.

Problem 6-4. Catching Cold: Exponential Spread of Disease

Exponential change, which we have encountered before in this book, is often a useful model for the spread of disease, especially as it first enters a population. The following shows how this model is derived.

Let $I(t)$ be the number of infected people at time t in a population N. Denote the transmission rate as α. During an increment of time dt we have $I(t)$ people infecting $\alpha I(t)$ people. Since a fraction of the total $(I(t)/N)$ are already infected, the number of new infections will be described by

$$\alpha I(t) - \alpha I(t)\left[\frac{I(t)}{N}\right] \qquad \text{6-1a}$$

or

$$\frac{dI(t)}{dt} = \alpha I(t)\left[1 - \frac{I(t)}{N}\right], \qquad \text{6-1b}$$

which is the basic *logistic* equation. An important observation is that early in the epidemic or when the population is large, $N \gg I(t)$, equation 6-1 reduces to the simple exponential, since

$$\frac{dI(t)}{I(t)} = \alpha dt; \qquad \text{6-2a}$$

therefore

$$I(t) = I(0)e^{\alpha t}. \qquad \text{6-2b}$$

Assume that a new mutation of the common cold appears, with an average transfer rate from one individual to the next of 0.08 per hour. How long will it take for this disease to spread from one initial carrier to one-quarter of the population of a town of 60,000? Be explicit about all assumptions. How do the assumptions become more problematic if you try to model spread of the cold to the entire population?

Solution 6-4

Using equation 6-2b to solve for the time at which one-quarter of the population is infected, $I(T) = 60,000/4 = 15,000$. α is given as 0.08/hour, and $I(0) = 1$. Plugging in the numbers,

$$15,000 = 1 \times e^{0.08T},$$

so

$$T = \ln(15,000)/0.08$$

$$= 120 \text{ hours (5 days)}.$$

The most problematic assumption in this case is that the average transfer time can hold for so many people with differing lifestyles. In any town, there are some people who come into contact with many people in a day, and others who only occasionally see others; if the initial carrier was part of the latter group, the initial rate could have been much lower. In addition, there are social groups of many types, and individuals will interact within the group much more often than outside of it. Consequently, while it might be reasonable to use the average rate for one-quarter of the population, the rate as the population becomes "saturated" might be very different.

Problem 6-I. Graphical Presentation

Graph the fraction of people affected over time as the population becomes saturated with the new cold virus. Compare this graph to figures 1-2 to 1-5. Are there any similarities? Discuss.

Problem 6-J. Small Groups

Now consider an elementary school classroom with 30 students and an incubation period (i.e., the amount of time from initial exposure to symptoms) for this virus of 40 hours. The chance of a student catching the cold from each other student who has it is 0.1 for each hour of contact, and the children are in the classroom together 6 hours per day.

a. Assuming that individual students can spread the disease to others as soon as they are infected, how long will it be until all of the children have the cold? Try to solve this problem both iteratively and analytically. What are the advantages and disadvantages of each approach?
b. Repeat (a) assuming some time lag between infection and contagion.

Problem 6-5. The Spread of AIDS: An Empirical Analysis, or, Does the Model Fit the Data?

The cumulative number of human immunodeficiency virus (HIV) and acquired immunodeficiency syndrome (AIDS) cases in the United States increased dramatically after the first documented cases in the early 1980s. The dynamic behavior of the growth of the epidemic is important because it informs the virology and the epidemiology of the disease, which in turn shape the medical options to treat individuals and the health policy options available to reduce the spread in the population.

In this problem you will build a pair of simple models for the growth of an epidemic, and then take a look at the assumptions in light of the observed spread of the AIDS virus.

a. Expected model:

    i. Build a simple model for the cumulative number of AIDS cases based only on the assumptions that AIDS is transmitted through sexual contact, that the transmission of AIDS is proportional to the number of sexual contacts, and that the behavior patterns of individuals do not change during the growth of the epidemic.

    ii. Discuss the rationale for choosing such a model.

b. Compare the data for the growth of AIDS shown in table 6-7 to your model. How good is the fit?

c. Consider instead an alternate representation of the actual AIDS data that are presented in table 6-7. Several researchers (Centers for Disease Control 1987; Colgate et al. 1989) have noticed that the data not only for total AIDS cases, but also for each of the major ethnic groups (Caucasian, black, Hispanic), as well as the regionally disaggregated totals, all follow a polynomial, not an exponential or other relationship.

    i. Using June 1980 as a reference (initial) date, identify a polynomial best-fit function for the data in table 6-7. To keep things simple, select the lowest-order polynomial model that fits the data well.

    ii. What does the model you found say about the rate of increase in the spread of the epidemic? In particular, what is the doubling time for this epidemic, and how does this compare to the doubling time for an exponential relationship?

d. Based on your answer to (a), discuss why this result is surprising. What does this tell you about the impact of public policies designed to combat the spread of AIDS?

Table 6-7 Cumulative Number of AIDS Cases

Year	Cumulative number of cases
to 1982	834
1983	2,901
1984	7,347
1985	15,552
1986	28,708
1987	49,851
1988	80,748
1989	115,748
1990	157,343
1991	199,169
1992	244,958
1993	347,245
1994	424,513
1995	395,663
1996	562,549

Source: U.S. Department of Commerce 1990, 1996.
Note: Some discrepancy may result from change in AIDS definition to be more inclusive of precursor illnesses.

Solution 6-5

Solution 6-5a(i)

In this case we build the model of spread of the disease on a proportionality between the number of people with AIDS, with some (unspecified) transmission coefficient. The transmission of AIDS in the model is based on sexual contacts, which we assume can be characterized by a probability: the number of sexual contacts per disease transmission. This is identical to the approach used in problem 6-4. Since α is unknown, this leads to a simple exponential, as in the last problem (equation 6-2b).

$$I(t) = I(0)e^{\alpha t}.$$

Solution 6-5a(ii)

An exponential model can only be a good fit when the number of infected people, $I(t)$, is small compared to N (which in this case is the population of the United States). From table 6-7 this appears safe, at least in the beginning of the epidemic. To increase from less than a thousand to half a million cases in only fourteen years is certainly frightening, however.

Solution 6-5b

This is a graphical and statistical exercise. The fit is not good. The growth in the early 1980s can be fitted reasonably well, but the takeoff more recently is too slow. Even though the goodness of fit R^2 value appears impressive (0.95), the exponential simply rises too rapidly. This already suggests that some other model may be more appropriate.

Solution 6-5c(i)

Using a spreadsheet best-fit command (we used the solver.exe add-on to Excel) yields a cubic equation of the form

$$N(t) = 174.6(t - t_0)^3 + 340, \qquad\qquad 6\text{-}3$$

where t_0 is the initial year (1980). This produces an exceptional fit ($R^2 = 0.996$). This is a good example of how R^2 can actually be deceptive: it only measures goodness of fit on an absolute scale. Normally you might be satisfied with $R^2 = 0.95$, but the early growth can be well represented by a number of models. In this case some sleuthing around with Excel or any other spreadsheet (and trying a range of curve fits, e.g., linear, quadratic, cubic, logarithmic, etc.) cannot only produce a better fit than did the exponential, but also provide a potentially important observation about the spread of AIDS. The relative quality of the exponential and cubic fits is shown in figures 6-2a and 6-2b.

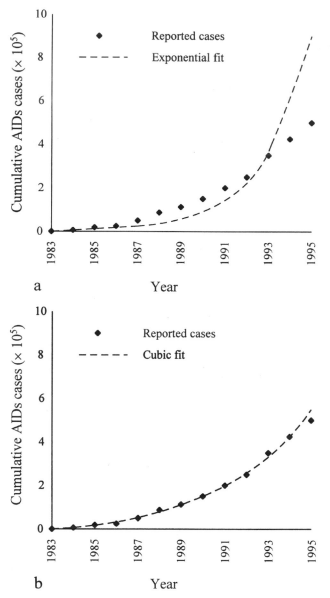

Figure 6-2. Cumulative number of reported AIDS cases in the United States from 1983 to 1996. Data from U.S. Department of Commerce 1990, 1996. (a) The number of AIDS cases as predicted by an exponential fit to the data ($R^2 = 0.95$) is overlaid. Notice the deviation beginning around 1992. (b) The number of AIDS cases as predicted by a cubic fit to the data ($R^2 = 0.996$) is overlaid.

This problem is now clearly an interesting "exception that validates the rule" in that we did not find the expected exponential, and since we did not, we have learned that something even more perplexing may be going on with the spread of AIDS. In the next section we will see further implications of the cubic growth equation.

Solution 6-5c(ii)

The rate of doubling in an exponential,

$$I(t) = I_0 e^n,$$

is easily found by setting $I(t) = I_0/2$, and finding that

$$\frac{I_0}{2} = I_0 \cdot e^{rt_{\text{doubling}}},$$

$$t_{\text{doubling}} = \frac{\ln(2)}{r} = \text{constant}.$$

Thus we find that for exponential growth, while doubling time does, of course, depend on the rate constant, it is also constant in time: for each time step there will be twice as many items as there were the previous time step. This leads to the famous problem of ecological management:

> A fish farmer looks out at his pond each day and notices that a weed is beginning to appear on the surface and cut off oxygen to the fish beneath. He watches it double in pond area covered each day until one day he notices that the pond is half covered. How much time does he have before the pond is totally covered?[1]

For the polynomial equation, where we use m instead of 3 for the more general solution,

$$I(t) = k_1 i^m,$$

for the simple case of $k_2 = 0$ we have

$$\left(\frac{dI(t)}{dt} \right) = k_1 m t^{(m-1)},$$

---

[1] The answer, of course, is one day!

so that the relative rate of growth, $[dI(t)/d(t)/I(t)]$, is equal to

$$\left(\frac{dI(t)}{dt}\right)\bigg/I(t) = \frac{k_1 m t^{(m-1)}}{k_1 t^m} = \frac{m}{t}.$$

Thus, for *any* type of polynomial growth (any value of m) we find that the relative growth rate *decreases* with time. That is, for any polynomial growth model, it takes longer and longer for the epidemic to double in size.

Solution 6-5d

The process we have followed in this problem has been to empirically critique simple models. The exponential model of part (*a*) would have seemed to make sense for AIDS, based on a truly simple transmission process: a new disease transmitted by the exchange of bodily fluids, which we can characterize by an exposure and transmission probability. The deviation from the exponential model is itself a surprise, given the simple and robust assumptions.

What does this imply for public health policy? There was a perception in the public health community that the increase in the doubling time of AIDS was due to the positive impacts of education programs that highlighted safe sex (the use of condoms) and the avoidance of high-risk, multiple-partner behaviors. It is possible, of course, that education and public health efforts did contribute to the slowing in the growth of the HIV/AIDS epidemic. However, in this problem we have found that the cubic growth model remained valid for at least the first decade (1980–1990) of the epidemic, back to a time before the public health response to this terrible disease was initiated.

We can learn more about the disease by looking at the curves in figure 6-3. Not only is the total epidemic showing cubic growth, but the pattern is the same in each of the constituent ethnic groups. Further, notice that each curve comes together in the early part of 1980 or 1981. This means that the disease was "seeded" in each population group at roughly the same time, and that AIDS was never a "gay," "straight," "minority," or "majority" disease, nor is it one confined to hemophiliacs or other groups. AIDS impacts all groups and has simply moved faster

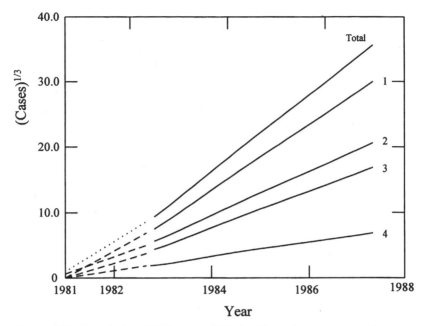

Figure 6-3. Cumulative AIDS cases plotted as the cube root versus time as reported by the Centers for Disease Control. After Colgate et al. 1989. 1, White; 2, Black; 3, Hispanic; 4, Unknown.

through communities that are either smaller, such as the homosexual community, or have far higher numbers of exposures, such as hemophiliacs, who initially had more exposures to infected "risky" blood than most people.

Problem 6-6. Double-Blind Study

The standard approach to determining the efficacy and safety of new medicines and treatments is the "double-blind" study, so called because neither the researchers nor the subjects know which individuals are getting the treatment and which are getting placebos.

Assume that a hypothetical researcher has decided to test whether a new medication will safely reduce teenage acne. Assume also that the medicine has been tested on animals and human skin cell cultures, with no known effects except for rare short-term rashes, which disappear

Table 6-8 Hypothetical Improvements in Skin Conditions for Treatment and Control Groups

Time (days)	Treatment group			Control group		
	No. of subjects	Average % improve-ment[a]	Range of improve-ment	No. of subjects	Average % improve-ment	Range of improve-ment
45	100	5	0–10	100	4	0–10
90	98	11	0–15	100	7	0–10
135	98	20	5–25	100	10	0–10
180	97	37	5–30	99	12	5–15
225	97	41	20–35	99	13	5–15
270	97	45	30–55	99	13	5–20
315	94	45	30–55	97	14	5–20
360	94	47	30–60	97	12	5–20

[a] % improvement is an aggregate measure that includes a severity and a surface area component. It is measured in increments of 5 percentage points.

rapidly after use is discontinued. She finds two hundred male teenagers who have serious but not severe acne (i.e., neither physically painful nor likely to cause permanent scarring if not treated), but have not responded to other treatments that are available.

To half of these (picked at random), she gives a cleanser with the new medicine, to half she gives the same cleanser without. She maintains a record of which patients receive which cleanser, but does not look at this record until after the experiment is completed. Hence the "double-blind": neither she nor the subjects knows who gets the treatment and who does not. She tells each to wash his face twice a day with the cleanser, and gives specific instructions on cleaning technique. Table 6-8 contains her results for the one-year duration of the study.

a. Discuss the similarities and differences between this and problem 6-3.
b. On the basis of these data, would you say that this medicine effectively treats teenage acne? Explain what you mean by "effective," and support your answer numerically. What potential problems do you see with your conclusions?

Solution 6-6

Double-blind studies, in which neither the researchers nor the subjects know who is receiving treatment and who is not, are the standard method used by drug companies and the Food and Drug Administration to evaluate new treatments and alternative uses for existing treatments. The double-blind feature ensures that no biases come into play, including subconscious ones. The goal of these tests is to determine whether the treatment is effective (i.e., has effects as intended) and safe (i.e., any adverse side effects are less severe than the beneficial effect).

Solution 6-6a

Both of these problems isolate an individual suspected causal agent and try to evaluate its effect given individuals who differ only in exposure to that agent. Methodologically, the fact that problem 6-3 looked at adverse effects and this one looks at beneficial effects is not a significant difference.

The major differences lie in measurement and number of individuals in the study. The Pliofilm study was able to include a very large number of individuals, a long time period, and a range of dose levels, all of which provide considerable statistical power. On the other hand, there are a number of potential problems that weaken the study, such as the accuracy of historical exposure estimates and disease diagnoses, and the validity of the case and control comparisons.

The double-blind case-control study, on the other hand, has much more precise measurement of exposure and effect, and the similarity between the test and control individuals is clear. However, the relatively small number of individuals reduces the statistical power, and the short test time necessarily ignores longer-term effects.

Solution 6-6b

It does appear that this treatment is effective in some sense. The problem postulated that none of the subjects had responded to conventional treatments, and the treatment group had a measured reduction in acne of about 50%. If the difference in average improvement for the treatment and control groups is statistically significant, the treatment is probably "effective." Calculating statistical significance should be second nature by now, so we will simply say the p-score is $\ll 0.01$.

There is still some uncertainty involved in this decision. Particularly important is that only "effective" has been addressed here. It is not evident whether there were no side effects, or if side effects were not recorded. Consequently, it is difficult to conclude that this treatment would be effective *and safe*. Second, there remain uncertainties in measurement, such as the meaning of and method for evaluating "severity and surface area," and the actual (as opposed to prescribed) application. Finally, this study only covers one year, and it is not clear what side effects might occur twenty years into the future.

Problem 6-K. Side Effects: Test for Safety

Table 6-9 indicates information on the only reported side effect: a temporary, and in some cases mildly itchy, rash. The researcher categorized the rash by number of individuals who reported rashes in each time period, with column A indicating visible rash only, and column B indicating visible and itchy rash. No cases of the rash were reported in the control group.

a. How does this affect your evaluation of "effective and safe?"
b. The company decides to try to market this medicine to "severe" acne sufferers, including those who respond somewhat to existing treatments. Design an experiment to test for this new use. Do the rashes reported in table 6-9 influence your design? Why and how? Why did the company not test this drug on "severe" acne sufferers first?

Table 6-9 Treatment Group Individuals Reporting Visible (A) and Itchy (B) Rashes

Time	Rashes		
(days)	A	B	Total
45	3	1	4
90	2	1	3
135	4	2	6
180	5	3	8
225	3	1	4
270	5	1	6
315	4	2	6
360	2	1	3

Problem 6-L. Low-Probability Effects

Assume that after several years, this drug became a common treatment for both serious and severe acne. Eventually, two individuals out of 154,543 users had potentially life-threatening allergic reactions, requiring hospitalization (neither died). Would you still consider this a "safe and effective" medication for

a. individuals like those in the original test group
b. the approximately 1,000 cases each year of severe acne that can't be treated with other medicines?

Note the similarity to Fen-Phen, where a treatment for severe obesity was given to many people who wanted to lose a few quick pounds. It later turned out that there was a fairly high incidence of heart valve problems associated with the drug, and it was quickly taken off the market.

Problem 6-M. Death by Cheese?

Suppose you found that 90% of people who died in a small town over a ten-year period had eaten cheese in the two weeks prior to their deaths. Should we be concerned about a threat from cheese? Why or why not? What data would we need to support a causal correlation between cheese consumption and death?

Problem 6-N. Cancer Clusters—Real or Not?

Suppose that in a community of 50,000 people, ten cases of "Leubkin's disease," a usually fatal form of cancer, occurred over a five-year period (note: this is not a real disease; we made it up for this problem). The annual incidence of Leubkin's in the population at large is estimated to be about two in a million.

a. Would you consider this a "Leubkin's cluster" or a statistical anomaly?

b. Suppose that there were five times as many Superfund sites in this community as in the rest of the country. Would this influence your answer to (a)? Why and how?

References

American Cancer Society. (1998). http://www.cancer.org/bottomcancinfo.html, accessed March 13, 1998.

Centers for Disease Control. (1987). *AIDS Public Information Data Set AIDS07*. Atlanta, GA: Centers for Disease Control.

Crump, K. (1996). "Risk of benzene-induced leukemia predicted from the Pliofilm cohort." *Environmental Health Perspectives* 104(6):1437–41.

Colgate, S. A., Stanley, E. A., Hyman, J. M., Layne, S. P., and Quallis, C. (1989). "Risk behavior-based model of the cubic growth of acquired immunodeficiency syndrome in the United States." *Proceedings of the National Academy of Sciences U.S.A.* 86:4793–97.

Harte, J., Holdren, C., Schneider, R., and Shirley, C. (1991). *Toxics A to Z: A Guide to Everyday Pollution Hazards*. Berkeley, CA: University of California Press.

U.S. Department of Commerce. (1990). *Statistical Abstract of the United States*. 110th ed. Washington, DC: U.S. Government Printing Office.

———. (1996). *Statistical Abstract of the United States*. 116th ed. Washington, DC: U.S. Government Printing Office.

7

Exposure Assessment

Introduction

In the previous chapters we have already given considerable attention to questions of the extent and the measurement of exposure. Consequently, the purpose of this chapter is to focus in on issues of special concern, such as individual versus aggregate exposure, sensitive receptors, and environmental justice concerns. It is also a useful context in which to further explore the use of Monte Carlo techniques, which are particularly salient in the case of exposure, where multiple models and data sets may be available.

Problem 7-1. Assessment of Exposures and Risks: The ChemLawn Claim

A few years ago, the lawn pesticide applicator company ChemLawn ran an advertising campaign in which the president of the company, referring to the product DMA 4, claimed that "a child would have to ingest twenty-five cups of treated lawn clippings to equal the toxicity of one baby aspirin." (Note: this statement was made before aspirin was associated with Reye's syndrome in children.) The company was eventually sued in New York State by the Attorney General and was forced to withdraw the advertisements. As an independent policy analyst, you can assess the validity of the statement.

DMA 4, one of the most popular herbicides for lawn applications in the United States, contains 2,4-D (as a dimethylamine salt, a nonvolatile form) as its only active ingredient. The product label calls for a standard application of two quarts per acre of lawn treated.

a. Assume that the ChemLawn person treats your lawn with DMA 4 at the label-specified dosage and that it is uniformly distributed. You then feed your child (weight 20 kg) twenty-five cups (English volume measure) of treated fresh grass immediately after the application, since you have a great deal of trust in advertising.

i. What is the concentration of 2,4-D per unit area of lawn?
ii. What is the concentration per unit weight of wet grass? Assume that only one-quarter of the chemical spray actually intercepts (sticks to) the grass.
iii. Assuming that the density of grass in the cup is ten times greater than the density of grass on the lawn, what is the density (weight per unit volume) of grass in the cup you are filling?
iv. How much 2,4-D is contained within each cup of grass?
v. What is the dose (per kg of body weight) delivered to your child from ingesting twenty-five cups of grass?
vi. What fraction of the rat LD_{50} dose does your child ingest? What fraction of the adult minimum lethal dose?

b. What fraction of the rat LD_{50} dose is the child exposed to when fed a single "baby" aspirin? What is the fraction of the adult lethal dose?

c. Using the results of (a) and (b) as crude indicators of the relative toxicity of the 2,4-D and aspirin exposures (knowledge of the slopes of the dose-response curves would be necessary for a more accurate determination), assess the accuracy of the ChemLawn advertisement if it is assumed that a human child's metabolism of these two chemicals would more closely resemble that of

i. A laboratory rat
ii. A human adult

d. Discuss briefly how variations in some of the parameters assumed for this calculation could affect the result. For instance, application on a windy day could cause significant spatial variations in concentration.

e. Whether or not the advertisement was accurate, do you think that it made a meaningful statement about product safety? Is acute toxicity the only relevant criterion for assessing the public health implications of these chemicals?

To work this problem you will need the following information:

- The rat LD_{50} dose for 2,4-D is 500 mg and for aspirin is 750 mg.

- The human adult minimum lethal dose (MLD) to two significant digits: For 2,4-D, it is poorly characterized (due to lack of information), but probably in the vicinity of 90 mg/kg (6.3 grams per 70 kg body weight). For aspirin (salicylic acid) it is 290 ~ 430 mg/kg (20–30 grams per 70 kg body weight). A child will become seriously ill and require hospitalization with a dose of around 100 mg/kg.
- Use a typical value for the wet weight of grasses per unit area of lawn, for example, 0.24 kg/m^2/cm of grass.
- Your grass is 5 cm tall at the time of spraying.
- The dosage of one "baby" aspirin is 81 mg/tablet.
- The fraction of chemical spray intercepted by vegetation is 1/4.

The following is an excerpt from a news story relating to this problem. A Lexis-Nexis search is sure to generate more articles.

Please Don't Eat the Daisies

Ah, springtime in suburbia: crocuses pop, daffodils bloom—and lawns sprout signs warning that they have been sprayed with pesticides so humans and other living things should keep off. These unlikely harbingers of the season are now required by law for professionally sprayed lawns in at least six states. Others are considering similar "posting" laws. At the same time, businesses are trying to convince consumers that lawn pesticides are almost as benign as a dandelion salad. "There is such flagrant misrepresentation consumers are unaware the products have risks," says Jay Feldman of the National Coalition Against the Misuse of Pesticides. New York Attorney General Robert Abrams concurs. He is suing ChemLawn, the largest lawn-care firm, alleging false and misleading safety claims in its ads.

Some 300 million pounds of pesticides are applied to lawns, gardens, parks and golf courses each year. Some of these compounds can cause nerve or kidney damage, sterility or cancer in lab animals and workers. But because epidemiological research is so sparse, it's nearly impossible to pin illnesses on lawn products. One closely watched attempt is the case of Navy Lt. George Prior. In 1982, according to his widow, he developed a headache, fever and nausea after golfing at Virginia's Army Navy Country Club, which was reportedly treated with the fungicide Daconil. He was suffering from toxic epidermal necrolysis (TEN), which makes skin fall off in sheets and causes organ failure. Prior died shortly after, and his widow contends that the TEN came from his exposure to chlorothalonil, an ingredient in Daconil. A lawsuit filed by Prior's widow is scheduled for trial this week. A spokesperson for Daconil's former manufacturer said the claim was "without merit."

Few people are as chemically sensitive as Prior. Yet lately there have been hints that healthy people are harmed by pesticides. "An increasing number who thought they had the flu are finding physicians can document a cause and effect relationship [to lawn chemicals]," says New York Assistant Attorney General Martha McCabe. For some, the risks are worse. In a controversial study, the National Cancer Institute found that Kansas farmers who apply 2,4-D, a weed killer used in 1,500 over-the-counter products, had a risk of non-Hodgkin's lymphoma, a rare cancer, up to seven times higher than average.

Now there is also concern over the solvents, dyes and other "inert" ingredients in pesticides. Of the 1,200 inerts, the Environmental Protection Agency considers 300 safe—and about 60 of "significant toxicological concern." Among them: benzene and chloroform. Both cause cancer in animals. Unlike active ingredients, inerts do not have to be listed on pesticide labels. But EPA has announced that the 60 most hazardous inerts must be removed from pesticides by October or be named on labels.

Clearly, killer lawns aren't sweeping suburbia. Roger Yeary, chief of health and safety at ChemLawn, argues that few homeowners get the "repeated exposure to a constant, low-level amount" of a chemical that produces chronic effects, since most lawn compounds quickly decompose. ChemLawn uses about 25 pesticides. In a brochure for customers, it stresses that none are "known or probable human carcinogens" and are "practically nontoxic." Yet one of its fungicides, mancozeb, breaks down into a compound EPA categorizes as a known human carcinogen. Another, chlorothalonil, causes cancer in animals. ChemLawn says it uses the fungicides hardly at all. Cancer, of course, is hardly the only concern. Data on the neurobehavioral effects of long-term, low-level exposure to lawn pesticides are poor or nonexistent.

Many consumers aren't aware of these doubts about safety. One major reason, according to a recent report by the U.S. General Accounting Office, is "false and misleading" safety claims by pesticide manufacturers. GAO's undercover team also found that professional applicators made unsupportable assertions, such as saying products "absolutely cannot harm children or pets." The New York suit cites ChemLawn's claim that a child would have to eat almost 10 cups of treated lawn clippings to equal the toxicity of one baby aspirin. In fact, the danger is not that people graze on the lawn but that they inhale fumes or absorb residues through skin.

Source: Copyright 1988 *Newsweek*, May 16, U.S. Edition. Section: Life/Style; Environment p. 76. Byline: Sharon Begley with Mary Hager in Washington.

Solution 7-1

Solution 7-1a (i)

The concentration of 2,4-D per unit area of lawn is

$$C = (2 \text{ quarts/acre}) \times (3.8 \text{ lb/gal}) \times (0.25 \text{ gal/quart})$$
$$\times (2.47 \text{ acre/hectare})$$
$$\times (0.455 \text{ kg/lb}),$$
$$= 2.1 \text{ kg/hectare} = 210 \text{ mg[2,4-D]/m}^2.$$

Solution 7-1a (ii)

The concentration per unit weight of grass is given by

$$[C \times (0.25)]/[((0.24 \text{ kg/m}^2) \times \text{cm}) \times (5 \text{ cm})]$$
$$- 44 \text{ mg[2,4-D]/kg.}$$

Solution 7-1a (iii)

The density of grass in the cup is ten times the density on the lawn, that is,

$$10 \times [[(1.2 \text{ kg/m}^2)]/(0.05 \text{ m})] = 240 \text{ kg/m}^3$$
$$\times 2.4 \times 10^{-4} \text{ kg/cm}^3.$$

Solution 7-1a (iv)

The quantity of 2,4-D ingested per cup is

$$(225 \text{ cm}^3/\text{cup}) \times (2.4 \times 10^{-4} \text{ kg/cm}^3) \times (44 \text{ mg[2,4-}D]/\text{kg})$$
$$= 2.4 \text{ mg[2,4-D]/cup.}$$

Solution 7-1a (v)

The dose from ingesting 25 cups (per kg of body weight) is

$$(25 \text{ cups}) \times (2.4 \text{ mg/cup})/(20 \text{ kg}) = 3 \text{ mg/kg}.$$

Solution 7-1a (vi)

The comparative doses are

$$\text{dose}/\text{LD}_{50} = 3 \text{ mg}/500 \text{ mg} = 0.006;$$
$$\text{dose}/\text{MLD} = 3 \text{ mg}/90 \text{ mg} = 0.033.$$

Therefore the ingested dose is about 0.6% of the LD_{50} and about 3.3% of the MLD for 2,4-D.

Solution 7-1b

The dose from one "baby" aspirin is

$$81 \text{ mg}/20 \text{ kg} = 4.05 \text{ mg/kg}.$$

The comparative doses are

$$\text{dose}/\text{LD}_{50} = 4 \text{ mg}/750 \text{ mg} = 0.0053;$$
$$\text{dose}/\text{MLD} = 4 \text{ mg}/290 \text{ mg} = 0.014.$$

Therefore the dose from one baby aspirin is about 0.54% of the LD_{50} and about 1.4% of the MLD.

Solution 7-1c

A quick comparison shows that, when using the LD_{50} to compare children to rats and the MLD to compare them to adults,

    i. The 2,4-D dose is 36% more toxic than the aspirin dose
    ii. The 2,4-D dose is 290% more toxic than the aspirin dose

Solution 7-1d

Factors that may affect the validity of the assumptions of the problem include differences in weather, uneven application, failure to understand the label and thus application of the wrong dosage, differences in how tightly you stuff the grass into the cups, and so on.

Solution 7-1e

A number of other points that we may choose to consider when assessing product safety are different pathways of exposure (i.e., not just through eating grass, but also through drinking water, dermal contact, etc.), the possibility of bioaccumulation, the effects of 2,4-D on the environment (e.g., animals and other vegetation), possible carcinogenic effects, and so on.

Problem 7-A. Can Adults and Children Be Treated the Same?

This problem has used the available data on 2,4-D, based on adult humans. Yet the exposure is to children. Suggest some differences that would make such a comparison problematic. Would it be reasonable to scale based on body weight? Why or why not?

Problem 7-2. Contaminated Milk

Consider an individual who drinks one carton of milk daily, purchasing it from a vending machine. Suppose a problem at the dairy leads to contamination of the milk with a pesticide to the extent of 4 mg per carton, 2 mg per carton, and 1 mg per carton on three successive days, before the problem is discovered and eliminated (whereafter the milk sold is again free of pesticide contamination). Let the retention function in the body for an initial intake, $Q(0)$, of this pesticide be given by $Q(t) = Q(0)e(-t/T)$, where t is measured in days following the intake and T is the pesticide's mean residence time in the body.

a. Generate an expression for the individual's body burden of the pesticide immediately after she has consumed her daily carton of milk on the third day of the contamination episode.
b. Generate an expression for the individual's body burden of the pesticide at any time subsequent to the consumption of her third and last carton of contaminated milk.
c. Sketch a graph of the individual's body burden of the pesticide from the time she drinks the first carton of milk until a month later, for three different assumptions about T:

  i. $T = 1$ day (very rapid excretion or decomposition)
  ii. $T = 10$ days (intermediate)
  iii. T = 100 days (slow excretion or decomposition)

Solution 7-2

Solution 7-2a

Again, the exponential is a useful tool for estimating some aspect of risk. Specifically, assuming instantaneous and complete uptake by the body, we can find an expression for $Q(3)$, the body burden on the day when the third dose is taken in. It is the sum of the first day's intake of 4 mg, diminished by two days of exponential decay; the second day's intake of 2 mg, diminished by one day of exponential decay; and the third day's intake of 1 mg; or

$$Q(3) = (4e^{(-2/T)} + 2e^{(-1/T)} + 1) \text{ mg.}$$

Solution 7-2b

For times subsequent to $t = 3$, the equation is rewritten with the initial body burden set at $Q(3)$, and time in days considered relative to the third day $(t - 3)$:

$$Q(t > 3) = Q(3)e^{[-(t-3)/T]} = [1 + 2e^{(-1/T)} + 4e^{(-2/T)}]e^{[-(t-3)/T]}.$$

Solution 7-2c

In this case, there are three discrete inputs, but the exponential equation can be generalized to include this. Denoting the intake dose at time t as $d(t)$, we find that the body burden on a given day is equal to the previous day's body burden decayed by one day, plus the new discrete input:

$$Q(t) = Q(t - 1)e^{(-1/T)} + d(t).$$

Putting this equation in a spreadsheet program allows us to compute the answer, which is summarized in table 7-1 and graphed in figure 7-1.

Problem 7-3. Biomass Fuels and Childhood Disease

This problem was contributed by Majid Ezzati. Traditional biomass fuels (dung, wood) are the main source of energy for most households in developing countries. In addition to their environmental impacts, traditional fuels result in high levels of indoor air pollution with adverse impacts on health. Pneumonia amongst children is one of the health impacts of indoor air pollution, constituting one of the most important causes of infant mortality. The use of less polluting fuels such as natural gas can reduce such adverse impacts.

About two billion people in the world use traditional biomass fuels for cooking. Biomass fuels account for 80% of all household fuel consumption. Using the following numbers calculate the decrease in infant mortality if all these people were to switch to natural gas for the

Table 7-1 Body Burdens Given Different Assumptions

					t	
T	0	1 +	2 +	3 +	10	20
1 day	0	4	3.47	2.28	2.1×10^{-3}	9.4×10^{-8}
10 day	0	4	5.62	6.09	3.02	1.11
100 day	0	4	5.96	6.90	6.43	5.82

Note: All values in mg.

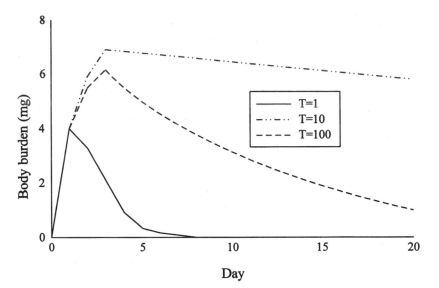

Figure 7-1. Graph of pesticide body burden over time as a function of T as described in problem 7-2.

source of energy for cooking (from Johansson et al. 1997):

People per household = 8 (assumption)
Wood use per household per day = 10 kg

Wood energy content = 420 ton/10^6 MJ
Natural gas = 32,000 m³/10^6 MJ

Solution 7-3

The equivalent amount of natural gas per family per day can be calculated to be

$$(32,000/420,000) \times 8 = 0.76 \text{ m}^3/\text{day}.$$

The emissions (suspended particulates) are

Fuelwood 3,800 μg/m³
Natural gas 1 μg/m³

Assume that cooking takes place for four hours each day and that the above emissions levels are the average over the cooking period (a standard assumption in suspended particulate measurements, as until recently we could only measure accumulated levels, so average emission was the best we had!). The average emission over the whole day can be calculated to be

$(4/24) \times 3,800 = 633 \ \mu g/m^3$ for fuelwood
0 for natural gas

To estimate health impacts, assume an increase in infant mortality due to pneumonia of $2.2/1000$ per $100 \ \mu g/m^3$ increase in suspended particulates (after controlling for income) (Smith 1996).

Then assuming four children in the household of whom one is an infant, the decrease in infant mortality will be

$$633/100 \times 2.2/1000 \times 1/8 \times (2 \times 10^9) = 3.5 \times 10^6.$$

Note that this deals only with risk of infant mortality due to pneumonia. Indoor air pollution from the use of biomass fuels for cooking is believed to contribute also to morbidity due to acute respiratory infection among children and adults (the exact relationship for the latter impacts is not quantified yet). Finally, wood collection often involves carrying large quantities of wood over long distances (often by women and children), contributing to backache.

Problem 7-4. Bioaccumulation of Heptachlor in Beef

For this problem, we consider the following *hypothetical* situation. From 1990 to 1994, heptachlor was used to control ants on pineapples in Hawaii (Harte et al. 1991). Clippings from these plants were used as cattle feed supplement before the normal delay time, which would allow the heptachlor to break down into harmless constituents. When samples of the clippings were tested in 1994, it was found that they contained between 20 and 100 ppm (95% confidence level, normally distributed) heptachlor (dry weight). Records showed that the clippings had been mixed with regular feed at a rate of 0 to 20% (uniformly distributed).

   a. Given the data below about exposure of beef and dairy cattle to heptachlor, calculate a point estimate of exposure to heptachlor of an average U.S. adult eating this beef. Calculate this first using average values and then using conservative estimates. Along with your answer, provide both measures and discussion of uncertainty.

b. Using your findings, estimate the additional lifetime cancer risk of an adult who ate beef at this rate for four years.

c. Use Monte Carlo analysis (we use Crystal Ball®) to recreate part (a) using the distributions given. Do the same for part (b), but suggest and justify (preferably document!) your distributions.

d. If you were the head of the U.S. Department of Agriculture, would you be concerned about your findings? Why or why not? For which parts of the calculation would you want more definite data? Are there other important exposure routes that may have been neglected in this calculation?

The additional information needed is listed below.

- The cattle feeding on this supplement were used for both milk and beef production, and were sold exclusively to local markets.
- These cattle ate 10–12 kg of feed per day (95% confidence interval, normal distribution).
- Adults eat 250–350 mg of beef per day (mean 270 and standard deviation 12, lognormal distribution).
- A ten-year-old child drinks 0.25–0.50 liters of milk per day (uniformly distributed).
- The people exposed to this beef and milk got it exclusively from this source.
- Relative bioavailability (RBA) of heptachlor to humans is 1.
- The cancer potency factor (CPF) for heptachlor is 4.5 (U.S. EPA, 1987) (in $(kg \times day)/mg_h$).
- Adult body weight is 40–100 kg (95% confidence interval, normal distribution) (after Finley 1994).
- Ten-year-old child body weight is 21.3–51.3 kg (95% confidence interval, normal distribution).
- Assume that beef on average contains 21.5% fat, and 25% is the 95th percentile value; milk on average contains 3.75% fat and (4.0% is the 95th percentile value.
- The biotransfer factor (BTF) for heptachlor to beef is 0.00655 (mg/kg fat)/(mg/day), and for heptachlor to milk is 0.0207 (mg/kg fat)/(mg/day).

Solution 7-4

This problem asks you to model a human health risk presented through foods. Heptachlor is a chlorinated organic pesticide that is highly restricted in the United States, but is still used in other parts of the world. Because it breaks down slowly in the soil, it is particularly useful against pests such as termites and ants (Harte et al. 1991). In addition to being found in beef through the path described in this problem, it has been found in significant concentrations in commercially harvested ocean fish.

As a class, organic pesticides tend to bioaccumulate in fatty tissues and are known as "lipophilic" compounds, reaching a "steady state" level in the body. Exposure to humans occurs when the fatty tissues (or vegetation with high oil content) are eaten; consequently, knowing the percentage of fat in the food is important.

Solving this type of problem involves three steps. The first is to model the bioaccumulation of the compound in food. The second is to estimate the dose that a person eating the food gets. The third is to estimate the effect that this dose will have. We do these steps sequentially in the solution. Be sure to keep in mind that the cancer potency doses are highly uncertain, as seen in the previous two chapters. Since the question asks for both average and conservative measures, we will do the two sets of calculations in parallel. In this case, the conservative calculation will use the ninty-fifth percentile level. Later problems will introduce point estimates with alternate levels of conservatism.

Solution 7-4a

Step 1: Bioaccumulation of Heptachlor in Beef. Daily heptachlor intake by cattle is calculated by multiplying total feed intake per day by the concentration of heptachlor in that food:

$$I_c = I_{feed} \times C_{feed} \times kg_{clippings}/kg_{feed}.$$

$$I_{c,\,average} = 11\,kg/day \times 60mg/kg_{feed} \times 0.1\,kg_{clippings}/kg_{feed}$$

$$= 66\,mg/day;$$

$$I_{c,\,average} = 12\,kg/day \times 100\,mg/kg_{feed} \times 0.2\,kg_{clippings}/kg_{feed}$$

$$= 240\,mg/day.$$

The steady-state bioaccumulation of heptachlor is calculated by multiplying the daily intake rate by the fat content and the biotransfer factor, which is determined either empirically or by estimation from the octanol-water partition coefficient, a standard measure of the extent to which the compound is lipophilic (Travis and Arms 1988).

$$Q_{beef} = I_c \times f_{beef} \times BTF_{beef}.$$

So

$$Q_{average} = 66 \, mg/d \times 0.215 \, kg_{fat}/kg_{beef}$$
$$\times 0.00655 \, (mg/kg_{fat})(d/mg)$$
$$= 0.0929 mg/kg_{beef};$$
$$Q_{conservative} = 240 \, mg/d \times 0.25 \, kg_{fat}/kg_{beef}$$
$$\times 0.00655 \, (mg/kg_{fat})(d/mg)$$
$$= 0.393 \, mg/kg_{beef}.$$

Step 2: Human Heptachlor Dose. Now that we have two point estimates for the heptachlor content of beef, we can calculate human intake from these two sources by multiplying beef consumption by the concentration of heptachlor in each:

$$I_h = Q_{beef} \times I_{beef}.$$

So the average dose is

$$I_{h,average} = 0.0929 \, mg/kg \times 0.30 \, kg/day$$
$$= 0.0279 \, mg/day,$$

and the conservative dose is

$$I_{h,conservative} = 0.393 \, mg/kg \times 0.35 \, kg/day$$
$$= 0.138 \, mg/day.$$

Given this level of intake over the four-year period, an average daily dose (ADD) can be calculated by assuming a seventy-year lifetime. The relative bioavailability is a measure of the fraction of the dose from food that is available to the body and consequently available to cause the mutations that lead to cancer.

$$\mathrm{ADD}_h(\text{life}) = (I_h \times \text{years of exposure}/\text{lifetime} \times \text{RBA})/$$
$$(\text{years in lifetime} \times \text{body weight}).$$

The average ADD is

$$\mathrm{ADD}_{h,\,\text{average}} = 0.0279\,\text{mg/d} \times 4\,\text{YpL} \times 1.0/(70\,\text{YinL} \times 90\text{kg})$$
$$= 1.77 \times 10^{-5}\,\text{mg/(kg} \times \text{d)}.$$

The conservative value is

$$\mathrm{ADD}_{h,\,\text{conservative}} = 0.138\,\text{mg/d} \times 4\text{YpL}$$
$$\times 1.0/(70\,\text{YinL} \times 110\text{kg})$$
$$= 7.17 \times 10^{-5}\,\text{mg/(kg} \times \text{d)}.$$

Solution 7-4b

Step 3: Carcinogen Risk. Using this point estimate for daily dose, averaged over an expected lifetime, the incremental lifetime cancer risk (the additional cancer risk attributable to this exposure) can be calculated by multiplying this dose by the cancer potency factor:

$$\mathrm{LCR} = \mathrm{ADD}_h \times \mathrm{CPF}.$$

Thus,

$$\mathrm{LCR}_{\text{average}} = 1.77 \times 10^{-5}\,\text{mg/(kg} \times \text{d)} \times 4.5\,(\text{kg} \times \text{d})/\text{mg}$$
$$= 8.0 \times 10^{-5},$$
$$\mathrm{LCR}_{\text{conservative}} = 7.16 \times 10^{-5}\,\text{mg/(kg} \times \text{d)} \times 4.5\,(\text{kg} \times \text{d})/\text{mg}$$
$$= 32 \times 10^{-5}.$$

Solution 7-4c

One-thousand Monte Carlo iterations yields data in table 7-2. Figures 7-2 and 7-3 show graphical outputs from the Monte Carlo runs.

Table 7-2 LCR and ADD Statistics Resulting from 1000 Monte Carlo Iterations

	Mean	Median	Standard deviation
LCR	7.2×10^{-5}	5.5×10^{-5}	6.8×10^{-5}
ADD	1.6×10^{-5}	1.2×10^{-5}	1.5×10^{-5}

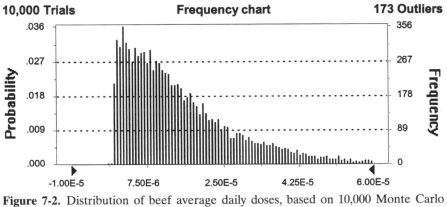

Figure 7-2. Distribution of beef average daily doses, based on 10,000 Monte Carlo iterations as described in problem 7-4c.

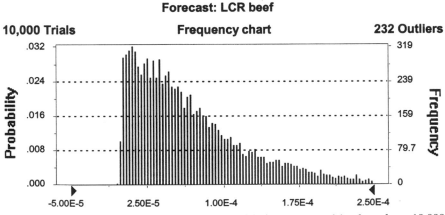
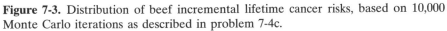

Figure 7-3. Distribution of beef incremental lifetime cancer risks, based on 10,000 Monte Carlo iterations as described in problem 7-4c.

Solution 7-4d

The current tolerated concentration of heptachlor in beef and milk is zero (0.0 μg/kg, Harte et al. 1991). Substances are often regulated in the United States when there is more than a 10^{-5} risk; again, even in the average value case here, that limit was exceeded.

One important exposure route left out is breast milk for women exposed to this beef and milk. Heptachlor accumulates in human milk much as it does in cows' milk, since the fat content of the two is quite similar (around 4%). Since its mother's milk is often the main diet for the first several months of an infant's life, this can mean a significant dose during that period.

Problem 7-B. Sensitive Receptors: Exposure to Children

Repeat this problem for exposure to heptachlor of a child who drank cows' milk contaminated as above from age six to ten years. In addition, discuss the implications of the exposed population size if the heptachlor cancer risks are based on a two-hit model, but one in ten thousand children have a prior mutation leading to an effective one-hit result.

Problem 7-C. Exposure via Breast Milk

Model exposure for an infant who drinks exclusively mother's milk for the first six months of life, assuming that the mother ate beef during this period as in the problem above. Clearly explain any assumptions and additional information you had to find.

Problem 7-5. Trichloroethylene Exposure at Woburn, Massachusetts

This problem was contributed by Dan Koznieczny. During the 1960s and 1970s, two wells that were subsequently found to be contaminated with the industrial solvent trichloroethylene (TCE) operated in the town of Woburn, Massachusetts. A number of the residents who were exposed to this water developed leukemia and sued two corporations who owned local factories in which the chemical was used. Jonathan Harr chronicled this lawsuit in *A Civil Action*, a compelling account that won the 1995 National Book Critics Circle Award (Harr 1995), as well as in a movie that was released in early 1999. Here, we explore exposure to TCE, which is commonly used to clean machinery and other metallic objects.

a. List the pathways in which the residents of a town that contains factories that use TCE could be exposed to the chemical.
b. Estimate the dose of TCE delivered to a child in Woburn, Massachusetts over a period of ten years while the wells were opened, using the well water concentrations and the outdoor air concentrations in table 7-3. Use the data in table 7-4 for 10–20 year olds. Calculate the dose using the most likely values for the parameters and Monte Carlo simulation.

Table 7-3 Concentration of TCE in Several Locations in Town with TCE Groundwater Concentration of 40 ppm

Location	Time (min)	Concentration (ppm)
Municipal Building, ladies rest room, water running	–	29
Home A, kitchen	0	< 0.4 (instrument detection level)
	19	26
	22	39
	27	55
	29	59
Home B, upstairs bathroom, shower on	0	< 0.4
	7	16
	10	29
	13	33
	17	66
Home B, downstairs bathroom, shower on	0	< 0.4
	9	52
	16	55

Source: Adelman 1985.

c. Use an LED_{10} curve fitting technique (see problem 5-2) to estimate TCE toxicity, based on the experimental results in table 7-5 involving Sprague-Dawley rats (Cotti and Perino 1985).

Table 7-4 Initial Assumptions

Duration of exposure	10 years
Level of TCE in drinking water	200 ppb
Age during exposure	10–20 years
Life span	70 years

Table 7-5 Toxicology Data for TCE Inhalation

Exposure (ppm)	Incidence of tumors	Number of rats alive at the time of the appearance of the first tumor
0	6	120
100	16	114
300	30	114
600	31	120

Note: The rats were exposed to this level of TCE in the atmospere for 7 hours daily, 5 days weekly, for 104 weeks. Then the rats were allowed to live out their natural life span.

Solution 7-5

Solution 7-5a

There are several specific pathways through which nonworkers could be exposed. First, TCE could reach residents through well water if the factory disposed of the solvent improperly. Drinking and cooking with the well water in a section of Woburn, Massachusetts, was indeed an exposure pathway for the residents of the town. The Massachusetts environmental protection agency ordered two Woburn wells shut down in 1979 because of measured concentrations of 267 parts per billion (ppb) in one (well G) and 183 ppb in the other (well H) (Harr 1995; see table 7-6).

Measurable concentrations of TCE can also be found in the outside air in some urban areas. One study (Andelman 1985) reported the range of mean outdoor air concentrations of TCE as 0.096–0.230 ppb in seven

Table 7-6 TCE Concentrations in Two Wells in Woburn, Massachusetts

Well	Year well opened	Concentration of TCE in 1979 (ppb)
G	1964	267
H	1967	183

U.S. cities (table 7-7). The maximum concentration measured during the experiment was 2.5 ppb, while the background concentration, estimated from a Pacific marine site, was 0.015 ppb. These are very small amounts of TCE, but it does represent a possible exposure pathway.

Another potential source of exposure is the inhalation of TCE that has evaporated from contaminated water. Andelman (1985) describes the results of an experiment to detect TCE in areas of a small community where TCE-contaminated water was run in the shower. The concentration of TCE in the water was 40 parts per million (ppm).

Solution 7-5b

There are three pathways to consider: inhalation from outside air, ingestion from well water, and inhalation from evaporated TCE from well water. Each estimate will require a number of assumptions, as discussed below.

Inhalation from Air. Here, we must estimate the concentration of TCE in the air and the amount of air that a typical resident inhaled from 1964 to 1979. The airborne concentration of TCE is by far the more difficult of the two factors to estimate, because we do not have any data for Woburn itself. Instead, we have average values for seven cities in the U.S. that range from 0.096 to 0.230 ppb. The maximum reading taken was 2.5 ppb. Let us take the midpoint of the range of averages as our first estimate, 0.163 ppb. With such limited data,

Table 7-7 Airbourne Concentrations of TCE in Seven U.S. Cities Over One-Week Period in 1980–1981 (ppb)

Range of means	Maximum	Background
0.096–0.230	2.5	0.015

assigning a distribution with any level of confidence is almost impossible. For the purpose of this problem, use a uniform distribution from 0.096 to 0.23 ppb.

Well-water Ingestion. Since we assume that we know the concentration of TCE in the well water, the only estimate that we need to make is how much water a typical Woburn citizen would ingest. The assumption that we know the concentration of TCE is quite a leap, because all we have is a single measurement for 1979. The well could have been contaminated with TCE since the beginning of its use, or since the week before the measurement was taken. Maybe there was a much higher concentration of TCE initially and it has decreased gradually over the years of well use. The data in table 7-8 represent the required distributions of tap-water intake and the inhalation rate (we'll need the inhalation rate data for the next section) from Finley et al. 1994.

The median tap-water intake rate is 0.87 l/day, which can serve as an initial point estimate. The range from the fifth to the ninty-fifth percentile, 0.24–2.7, provides a rough idea of the distribution of the data. For the most accurate results the distribution itself can be used. You might be wondering how to enter these distributions into your Monte Carlo program. If your program does not accept cumulative probability density distributions, you might have to manually fit one of the distributions your program can accept to the data. Table 7-9 contains distributions and parameters that Crystal Ball can accept for the two data sets above.

Table 7-8 Useful Distributions

Percentile	Tapwater intake (l / day) (ages 11–18)	Inhalation rate (m³ / day) (ages 10–18)
5th	0.24	9.1
10th	0.35	9.8
25th	0.57	11.2
50th	0.87	13.1
75th	1.2	15.3
90th	1.7	17.7
95th	2	19.3
99th	2.7	22.5

Table 7-9 Approximate Distributions for Use with Crystal Ball

Value	Distribution	Parameters	Error (sum of squares)
Tapwater ingestion rate (l/day)	extreme value	mode = 0.706, scale = 0.421	5.7×10^{-5}
Inhalation rate (m^3/day)	extreme value	mode = 12.05, scale = 2.60	5.7×10^{-5}

Inhalation of Evaporated TCE. Research by Andelman (1985) suggests that this pathway could be significant for estimating TCE exposure, but many assumptions will be necessary in order to convert the information in table 7-9 above to a dose estimate. Some of these assumptions are listed below:

- A typical adult takes one shower a day, lasting 10 minutes. The distribution of shower time can be approximated by a normal distribution with 10 minutes as the mean value and a standard deviation of 5 minutes. One can cut off the distribution at 3 minutes to avoid the complication of negative shower times!

- The distribution of inhalation rates for individuals of the appropriate age is as shown in table 7-8.

- During a shower, the concentration of TCE in the room could reach (from the data above) 26, 29, or 52 ppm if the concentration of TCE in the groundwater was 40 ppm. Let's average for a point estimate of 36 ppm with a range of 26–52 ppm. Now, we have to adjust for the lower TCE groundwater concentrations in Woburn. Assuming that the relationship between air and water concentrations is linear, the ratio between the tests described above and the Woburn concentration would be 40:0.200. So the corrected concentrations are an average of 180 ppb and a range 130–260 ppb. We use a uniform distribution ranging from 0 to 300 ppb for the concentration of TCE in the shower.

- Other activities involving running water, such as cooking, washing dishes, and watering the lawn, are performed in better-ventilated areas than the shower, and the TCE concentration from evaporation from water is negligible.

Table 7-10 Concentration Due to TCE Evaporation in Shower

Quantity	Distribution	Parameters
Concentration of TCE in shower	Uniform	Range = 0–300 ppb
Time spent in shower	Normal	Mean = 10 min
		SD = 5 min
		Cutoff = 3 min

Table 7-10 summarizes the quantitative estimates we have made.

Based on the data for TCE concentrations in showers, it is a reasonable assumption that the time taken for a shower and the concentration of TCE are correlated with each other. That is, slow showers correspond to high concentrations of TCE more often than fast showers do. A Monte Carlo model of this problem should take this factor into account.

Having estimated all of these values, we are finally ready to estimate the average daily dose that a 10–20-year-old individual would receive over a ten year period of exposure. The actual calculation takes the form:

$$\text{dose} = \frac{(\text{TCE concentration in medium}) \times (\text{daily consumption of medium})}{(\text{body weight})}.$$

The last piece of the puzzle we need is an estimate of the body weight of a 10–20 year old. We take this to be normally distributed around a mean value of 52 kg with a standard deviation of 10 kg. With this information, we can solve the problem; table 7-11 summarizes the results.

Solution 7-5c

As in chapter 5, the steps required to complete this problem can be listed:

1. calculate the average lifetime dose of TCE delivered to the rats;
2. convert the rat dose to an estimated human dose;
3. determine 95% confidence intervals for each data point; and

Table 7-11 Results

Quantity	Most likely values	Monte Carlo simulation (95% confidence upper limit)	Monte Carlo, TCE concentration and shower time correlated
Dose from ingestion	4.8×10^{-4}	5.41×10^{-4} (1.12×10^{-3})	5.41×10^{-4} (1.17×10^{-3})
Dose from outside air	7.8×10^{-6}	7.75×10^{-6} (1.38×10^{-5})	7.72×10^{-6} (1.35×10^{-5})
Dose from shower	4.8×10^{-5}	5.34×10^{-5} (1.40×10^{-4})	6.48×10^{-5} (1.82×10^{-4})
Combined inhalation dose	5.5×10^{-5}	6.12×10^{-5} (1.51×10^{-4})	7.25×10^{-5} (1.92×10^{-4})

4. fit a curve through the upper limits of the data and the origin and estimate the dose resulting in a 10% increased risk of cancer.

The data on laboratory rats needed to do the calculation are provided in table 7-12.

The average lifetime rat dose is

$$\text{dose} = \frac{\text{TCE conc.} \times \text{inhalation rate}}{\text{body weight}} \times \text{fraction of lifetime exposed.}$$

We convert the rat dose, as given in table 7-13, to an estimated human dose by multiplying by the scaling factor found in chapter 5: $(BW_{human}/BW_{rat})^{0.25}$. Estimated human doses found using the average value for human body weight (ages 10–20) calculated above (52 kg) are given in table 7-14.

The 95% confidence interval can be calculated from the cancer risk and the number of trials (the number of rats per data point) by assuming that each data point consists of a binomial distribution with a

Table 7-12 Average Rat Data

Inhalation rate	$0.11 \text{ m}^3/\text{day}$
Body weight	0.35 kg
Life span	2.5 years

Table 7-13 Average Lifetime Dose for Rat

Average lifetime dose (mg/(kg-day))	Fraction of rats developing tumor (cancer risk)
0	0.05
6.4	0.14
19.2	0.26
38.5	0.26

Table 7-14 Estimated Human Dose

Average lifetime dose (mg/(kg-day))	Cancer risk
0	0.05
19.0	0.14
57.0	0.26
114.1	0.26

probability equal to the cancer risk. The confidence interval is (cancer risk) $\pm 2 \times \sqrt{[np(1 - p)]}/n$ (table 7-15).

Fitting a third-order polynomial to the data above and the origin yields an equation:

$$y = 7 \times 10^{-7}x^3 - 0.0002x^2 + 0.014x + 0.$$

According to the is equation, the estimated dose leading to an increased cancer risk of 10% is

$$LED_{10} = 8.0 \text{ mg/(kg-day)}.$$

Table 7-15 Upper Limit of Cancer Risk

Dose (mg/(kg-day))	Cancer risk	95% interval	Upper limit cancer risk
0	0.05	0.04	0.09
19.0	0.14	0.065	0.21
57.0	0.26	0.08	0.35
114.1	0.26	0.08	0.34

Solution 7-5d

This part is easy, given our assumptions:

$$\text{increased cancer risk} = \text{dose} \times 0.1/\text{LED}_{10}.$$

The results are shown in table 7-16. Unfortunately, it is difficult to combine an ingestion and an inhalation dose because TCE could have very different health effects depending on the exposure pathway. The amount of TCE that actually enters the body through the lungs could be different than the amount that enters through the intestinal tract. Also, the relative concentrations of TCE in various organs of the body could be different, causing different effects for each pathway. It should also be noted that the result we arrived at depends on too many arbitrary assumptions to be useful in determining what actually happened in Woburn, Massachusetts. Still, this problem provides some insight into what data it would be important to collect and how a more complete data set would be used to determine whether a link could be established between an apparent leukemia cluster in Woburn and the concentration of TCE in the wells.

_____

Problem 7-D. TCE at Woburn: The Big Picture

This problem will take a considerable amount of time and research, and may be more appropriate as a class or semester project than an individual problem. Despite the inconclusive results from problem 7-5 using the given methods and data, there is nonetheless considerable evidence of a link between the contaminated wells and the childhood leukemia cluster at Woburn. Consult the material and references in Harr 1995, as well as other information on TCE, leukemia, and cancer clusters, for which there is a considerable literature. Use this informa-

Table 7-16 Results

Most likely values	6.7×10^{-6}
Monte Carlo (95% upper limit)	7.5×10^{-6} (1.53×10^{-5})
Monte Carlo with correlation	7.6×10^{-6} (1.57×10^{-5})

tion to construct evidence on the hypothesized link between TCE and leukemia at Woburn. Make an argument for or against such a link. Discuss the data and methods that you use and the types and extent of uncertainty in your analysis, and suggest what additional data and analysis might provide better insight into the situation.

Problem 7-E. Probability Distribution for Radon Exposures (or Risks)

This problem was contributed by Tony Nero. The distribution of (annual-average living-area) radon concentrations in single-family homes is adequately represented by a lognormal distribution with a median (or geometric mean) of 0.9 pCi/l and a geometric standard deviation (GSD) of 2.8. However, due to the fact that people live in more (sometimes many more) than one home during their lives, the distribution of lifetime exposures—in units of (pCi/l)-year—is significantly different than that obtained by simply multiplying the concentration distribution by seventy years or so.

a. Since people are said to move on the average of about every 7 years, use the simplest possible set of assumptions to develop a lifetime exposure distribution. Assume that every person lives in 10 different houses during his/her lifetime, in each case for 7 years, and that each move is to a random house in the national concentration distribution. How does the average exposure, and the percentage above 280 and 400 (pCi/l)-year, compare with results assuming each person lives his/her entire 70-year lifetime in a single house?

b. Go to the next level of sophistication by constructing a moving pattern where each person moves from house to house after a number of years drawn from an appropriate random distribution (for example, a uniform or triangular distribution that spans the main range of periods that people tend to live in the same house). How do the resulting average and percentages compare with those from part a?

c. Now try to account for the fact that different groups of the population tend to have different moving patterns, some moving more frequently, and others significantly less frequently than the average person. Does this factor make much difference in the percentages exceeding the lifetime exposures as calculated in (b)?

Different regions of the country have different distributions of radon concentration. At the same time, a substantial fraction of moves are to other homes in the same region; moves to another region are much less frequent.

d. Suggest how you might account for this factor.

e. Repeat problem 7-E, assuming there are four or five different regions with different medians, but probably about the same GSD.

Problem 7-F. PBPK Models and Gender Differences in the Uptake of Benzene

This problem uses a simple pharmacokinetic model to examine gender differences in the uptake of benzene by the body. Consider only the simplified system depicted in figure 7-4. Inhaled benzene is absorbed by the blood, then distributed to the fat and the liver. Both the fat and the liver may release benzene back into the blood flow, and some benzene

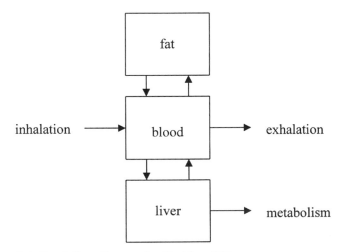

Figure 7-4. Stock-flow diagram for Problem 7-F.

is metabolized in the liver. The blood may release benzene back into the air through exhalation.

The distribution of benzene from air to blood, blood to fat, fat to blood, and so on, may be calculated using the values in table 7-17 for cardiac output, lung ventilation, blood flow fractions, tissue volume fractions, partitioning fractions, and metabolic constant. Assume the following:

- Partitioning fractions describe the distribution of benzene between substances (air and blood, blood and fat, blood and liver), accounting for both the relative uptake of each substance and the difference in density (all concentrations are in mg/l)
- The release of benzene from an organ (to the veins) is proportional to the tissue volume fraction of the organ
- The metabolic constant multiplied by the liver concentration gives the total metabolism
- The uptake of benzene by an organ (from the arteries) is proportional to the blood flow fraction to the organ

  a. Write out the system of equations describing the model depicted in figure 7-4.
  b. Solve the system for the steady-state concentration of benzene in the blood. Calculate the steady-state concentrations for men and women using the values in table 7-17. Assume a metabolic

Table 7-17 PBPK Model Parameters for Problem 7-F

Parameter	Men	Women
Lung ventilation (liters/hr)	450.00	363.00
Cardiac output (liters/hr)	336.00	288.00
Blood flow fractions (%)		
to liver	25.00	25.00
to fat	8.00	8.00
Tissue volume fractions (%)		
liver	2.60	2.30
fat	20.00	30.00
Partitioning fractions		
blood:air	0.886: 0.114	0.891: 0.109
liver:blood	0.747: 0.253	0.737: 0.263
fat:blood	0.982: 0.018	0.981: 0.019

constant of 12 liters/hour, and an air benzene concentration of 25 ppmv (0.0825 mg/l).

   c. Discuss the sensitivity of the steady-state blood benzene concentration to each parameter. Which parameters exert the most control over blood concentration? How would you explain the observed difference in male and female benzene blood concentrations?

Problem 7-G. Acme Landfill

In 1992, Thompson et al. developed the following scenario to demonstrate how Monte Carlo modeling can be used to incorporate "extended uncertainty analyses in public health risk assessment."

Acme, a private company, owns the 500×600 foot site which is located at the edge of Central City. Beginning its operation in the early 1850s, Acme used and maintained 27 coke ovens and two gas holding tanks, and produced blue gas at the site until 1945, when the buildings and equipment were demolished. From 1952 to 1988, Baker Company leased the southern third of the property from Acme for use as a fuel storage and tank truck depot.

Central City created a twenty-acre City Park to the north of the site in 1933. In 1989, Central City asked Acme to donate or sell the whole property to them to enlarge the City Park. At first, thinking that they might develop the site, Acme cleared the site and removed the visually stained surface soils. However, in further talks with the city last year, Acme agreed in principle to sell the property for inclusion in the park. Depending on the outcome of a site risk assessment for the surface soils on the site, Acme retains the right to limit the use of the site to activities with little or no soil contact (e.g., a parking lot with concession stands or a swimming pool with a large concrete pavilion).

   a. Using the following assumptions and data, generate two LCR (lifetime cancer risk) analyses algebraically, one using expected values as point estimates, the second employing the standard EPA conservative point estimate approach to find reasonable maximum exposure (RME).

   b. Repeat the above exercise using Monte Carlo analysis. Compare your findings in part (a), and discuss the merits and shortcomings of each approach.

c. Suggest possible distributions for the parameters that are only given as point estimates. Be sure to include a sensitivity analysis.

d. Discuss whether including this area in a park is a sensible choice for Central City. If you were making this decision for Acme, would you require the low-soil-contact options, or would you allow a playground? Why?

Data

Table 7-18 contains the data needed to answer this problem. Column 1 contains the name and symbol for each parameter used in the model. In column 2 we list the units for each parameter. Column 3 tells whether the data are used for modeling skin (dermal) contact with soil, ingestion of soil, or both. Column 4 shows the expected (mean) value for each parameter. Column 5 shows the point estimate values that Thompson et al. would use to calculate the reasonable maximum exposure (RME) according to EPA's *Interim Final Human Health Evaluation Manual* (U.S. EPA 1989). The final column shows the distribution that is entered into a Monte Carlo model. These distributions were chosen by Thompson et al. to reflect their best estimates, and to reflect the best available information from the literature. (Note: this data table can be downloaded from our website, http://socrates.berkeley.edu/erg/swri.)

Model

The following equations are used to calculate the lifetime cancer risk (LCR), which is the "additional probability that a person will develop cancer during the lifetime in which exposure occurs." To find the LCR, the doses received through dermal and oral contact are averaged over the individual's expected lifetime to get an average daily dose (ADD), which is then multiplied by a cancer potency factor (CPF) derived from epidemiological, toxicological, and other studies.

Equation 7-1 is the soil ingestion model used to find ADD(life), or average daily dose of a compound, averaged over the life during which

Table 7-18 Information for Problem 7-F Parameters

Name, symbol	Units	Model	Expected value	RME point estimate	Distribution
Average body weight, BW	kg	both	47	47[a]	normal (47, 8.3)
Time soil stays on skin, T	hr	dermal	6	8[b]	normal (6, 1)
Average body surface area	m^2	dermal	1.4	1.4[a]	normal (1.4, 0.17)
Fraction skin area exposed, BF		dermal	0.12	0.2[c]	lognormal (−2.15, 0.5)
Skin soil loading, SL	mg/c^2	dermal	1	1.0[a]	uniform (0.75, 1.25)
Soil ingestion rate, SIngR	mg/d	ingestion	31	50[d]	lognormal (3.44, 0.8)
Exposure days per week, DpW	d/wk	both	1	1	(none)
Exposure weeks per year, WpY	wk/yr	both	20	20	(none)
Exposure years per life, YpL	yr/life	both	10	10	(none)
Days in year, DinY	d/yr	both	364	364	(none)
Years in lifetime, YinL	yr/life	both	70	70	(none)
Soil properties					
Soil bulk density, ρ_b	kg/m^3	dermal	1600	1600[a]	normal (1600, 80)
Soil porosity, ϕ	m^3/m^3	dermal	0.5	0.5	(none)
Soil water content, θ	m^3/m^3	dermal	0.3	0.3	(none)
Organic carbon fraction, f_{oc}		dermal	0.02	0.02	(none)
Human skin properties					
Skin thickness, δ_{skin}	m	dermal	1.50×10^{-5}	1.5×10^{-5}	(none)
Skin fat content, f_{fat}	kg/kg	dermal	0.1	0.1	(none)
Skin water content, γ	m^3/m^3	dermal	0.3	0.3[a]	normal (0.3, 0.05)
Boundary layer size, δ_a	m	dermal	0.0045	0.0045	(none)
K_{ow}, benzene		dermal	135	135	(none)
K_h, benzene		dermal	0.224	0.224	(none)
K_{ow}, BaP		dermal	1.55×10^6	1.55×10^6	(none)
K_h, BaP		dermal	2.04×10^{-5}	2.04×10^{-5}	(none)
D_{air}	m^2/s	dermal	5.00×10^{-6}	5×10^{-6}	(none)
D_{water}	m^2/s	dermal	5.00×10^{-10}	5×10^{-10}	(none)

Table 7-18 (*Continued*)

Name, symbol	Units	Model	Expected value	RME point estimate	Distribution
Soil concentrations					
$Cs_{benzene}$	mg/kg	both	2.31	3.39[e]	lognormal (0.84, 0.77)
Cs_{BaP}	mg/kg	both	16.6	29.49[e]	lognormal (2.81, 0.68)
Relative bioavailabilities					
RBA, benzene		ingestion	1	1	(none)
RBA, BaP		ingestion	0.3	0.3	(none)
Cancer potency factors					
$CPF_{benzene}$	(kg-d)/mg	both	1.32×10^{-2}	2.9×10^{-2}[f]	lognormal $(-4.33, 0.67)$
CPF_{BaP}	(kg-d)/mg	both	0.45	11.5[g]	lognormal $(-0.79, 2.39)$

[a] Mean.
[b] 95th percentile.
[c] 85th percentile.
[d] 72nd percentile.
[e] 95th percentile C.I. of mean.
[f] 88th percentile.
[g] 91st percentile.

exposure occurs, in units of mg/(kg-day). This equation takes into account

- the concentration of the chemical in the soil;
- the amount of soil the hypothetical individual ingests;
- the body weight of the individual (a given dose will have a larger impact on a small person than on a large person);
- the life expectancy of the hypothetical individual (usually taken to be seventy years); and
- a "relative bioavailability," which adjusts for the extent to which the chemical simply passes through the body, or remains "available" in the body to cause mutations.

$$\text{ADD(life)} = \frac{Cs \cdot SIngR \cdot RBA \cdot DpW \cdot WpY \cdot YpL \cdot 10^{-6}\,\text{kg/mg}}{BW \cdot DinY \cdot YinL}.$$

7-1

Equations 7-2, 7-3, 7-4, and 7-5 represent the model used to find the ADD(life) for dermal contact with soil. This model accounts for the same factors as in the ingestion model. However, chemical exposure through the skin is a more complex pathway. The personal exposure factor (PEF) is based on a one-time deposition model published in 1990 by McKone to estimate dermal uptake (McKone 1990). The PEF incorporates properties of the soil (bulk density, porosity, water content, and organic carbon content), human skin (thickness, fat content, water content, and boundary layer size), and chemicals (octane/water partition ratio (K_{ow}), and Henry's constant (K_h) of air and water).

$$\text{ADD(life)} = \frac{\text{Cs} \cdot \text{PEF} \cdot \text{DpW} \cdot \text{WpY} \cdot \text{YpL}}{\text{DinY} \cdot \text{YinL}}, \qquad 7\text{-}2$$

where

$$\text{PEF} = \frac{\text{SL} \cdot \text{BF} \cdot \text{SA} \cdot 0.01}{\text{BW}} \left(\frac{K_u}{K_u + K_v} \right)$$

$$\times \left(1 - \exp\left(-\frac{3600(\rho_b + 1000\theta + \phi - \theta)(K_u + K_v)T}{\text{SL} \cdot 0.01} \right) \right), \qquad 7\text{-}3$$

$$K_v = \frac{0.000005 \cdot K_h}{\delta_a(4.8 \times 10^{-4}\rho_b f_{oc} K_{ow} + \theta + K_h(\phi - \theta))}, \qquad 7\text{-}4$$

$$\frac{1}{K_u} = \frac{\delta_{\text{skin}} f_{\text{fat}} K_{ow}}{D_{\text{water}} \gamma^{4/3}}$$

$$+ \frac{\text{SL} \cdot 0.01 \cdot \phi^2 (4.8 \times 10^{-4}\rho_b f_{oc} K_{ow} + \theta + K_h(\phi - \theta))}{(\rho_b + 1000 \cdot \theta + \phi - \theta)\left((\phi - \theta)^{3.33} D_{\text{air}} K_h + \theta^{3.33} D_{\text{water}}\right)}. \qquad 7\text{-}5$$

The lifetime cancer risk for each chemical is found by summing the two ADDs (dermal and ingestion) for each chemical and multiplying by the appropriate cancer potency factor.

References

Andelman, J. B. (1985). "Human exposures to volatile halogenated organic compounds in indoor and outdoor air." *Environmental Health Perspectives* 62:313–318.

Cotti, G. and Perino, G. (1985). "Long-term carcinogenicity bioassays on trichloroethylene administered by inhalation to Sprague–Dawley rats and Swiss mice and B6C3F1 mice." *Annals of the New York Academy of Sciences* 534:316–42.

Finley, B., Proctor, D., Scott, P., Harrington, N., Paustenbach, D., and Price, P. (1994), "Recommended distributions for exposure factors frequently used in health risk assessment." *Risk Analysis* 14(4):533–53.

Harr, J. (1995). *A Civil Action*. New York, NY: Random House (Vintage Books).

Harte, J., Holdren, C., Schneider, R., and Shirley, C. (1991). *Toxics A to Z*. Berkeley, CA: University of California Press.

Johansson, T. B., Reddy, A. K. N., and Williams, R. H. (1987). *Energy After Rio*. New York: United Nations Publications.

McKone, T. E. (1990). "Dermal uptake of organic chemicals from a soil matrix." *Risk Analysis* 10(3):407–419.

Smith, K. R. (1996). "Indoor air pollution in developing countries: Growing evidence of its role in the global disease burden." Invited address. Vol. 3, pp. 33–44 in Ikeda, K. and Iwata, T. (Eds.) *Indoor Air '96, The 7th International Conference on Indoor Air Quality and Climate* Tokyo: Institute of Public Health.

Thompson, K. M., Burmaster, D. E., and Crouch, E. A. (1992). "Monte Carlo techniques for quantitative uncertainty analysis in public health risk assessments," *Risk Analysis* 12(1):53–63.

Travis, C. C. and Arms, A. D. (1988). "Bioconcentration of organics in beef, milk and vegetation." *Environmental Science and Technology*. 22(3): 271–274.

U.S. EPA. (1989). *Risk Assessment Guidance for Superfund, Volume 1, Human Health Evaluation Manual, Part A, Interim Final*. EPA/540/1-89/002. Washington, DC: Office of Emergency and Remedial Response.

8

Technological Risk

Introduction

As we have explored in the preceding chapters, risk analysis tools are applicable to a broad range of problems and situations in fields ranging from individual and public health policy analysis to forecasting everything from national and household energy consumption to population growth to economic and financial modeling. Risk analysis in various forms has historically been applied when decision alternatives are evaluated. Probabilistic risk analysis has a very long history of both implementation and influence, dating back at least to 3200 B.C. and the Mesopotamian Asipu priests who dispensed advice based on a risk-balancing methodology (Oppenheim 1977; Grier 1981; Covello and Mumpower 1985).

Risk analysis emerged as a discipline of operations and decision engineering, which focuses on the evaluation of technological systems. For instance, the risks of various forms of transportation—walking, bicycle riding, driving a car, taking a train, flying, and space flight—lend themselves to an engineering analysis, and then naturally to comparisons of risk probabilities. The accident rates for each type of travel are available, as is information on a variety of technological failure modes: probability of tire blowout or collision per bicycle-mile traveled; probability of brake failure on cars; airplane damage from a range of weather conditions; catastrophic explosion during space shuttle launch, and so forth. Technical risks can also "cascade": damage to the electrical system on an airplane or unmanned satellite can impact other systems, crippling the vehicle. Cascades can be of any size, from a single mechanical system to global scale.

In a true nightmare scenario, during 1942 discussions of the feasibility of fission and fusion weapons (risks in themselves), Edward Teller considered the possibility that the high temperature of a nuclear explosion might initiate a fusion chain-reaction of the hydrogen in the Earth's atmosphere or oceans, burning up the world (Compton 1956).

The story of the analysis undertaken to evaluate this risk is fascinating in itself (Dudley 1975; Kammen et al. 1994), but the conclusion was that it is essentially impossible to ignite and sustain the hydrogen reaction in the atmosphere or the oceans.

The appeal of technical risk analysis, particularly in the 1950s and 1960s as engineering analysis of probabilistic failure rates evolved (Starr 1969), was that it became possible to quantify some beneficial as well as potentially negative impacts of technological systems. In this chapter we explore a variety of methods and applications of technical risk analysis, and examine both the power and the limitations of a quantitative approach to systems analysis. The risks associated with various transportation systems, for example, are clearly more complex than just the accident probabilities and consequences. As we shall examine in the chapters on decision making (9) and risk perception and communication (10), the degree of personal control, convenience, and familiarity associated with cars versus buses or trains leads to very different individual reactions to risk. In the end, risk management combines technical risk analysis and social valuation.

Problem 8-1. Lethality of Plutonium

This problem, while not derived from any specific situation, indicates the extent to which a very small amount of plutonium (Pu-239) can create substantial risks. Six kilograms of Pu-239 is approximately the amount used in the bomb dropped on Nagasaki. At about four times the atomic weight of iron, 6 kg of Pu-239 is about the size of a baseball, yet the nonexplosive (i.e., radiation) impacts of that amount, even when spread widely, can cause devastating damage to a large area.

Plutonium-239 decays initially by alpha particle emission with a half-life of 24,100 years, releasing 6.2×10^{-2} Ci/g_{Pu239}. One curie (Ci) is equivalent to 3.7×10^{10} becquerels (Bq) where one Bq is one nuclear transformation (decay) per second. Alpha particle emitters are hazardous to human health because they cause direct damage to DNA, thereby promoting potentially tumorogenic mutations. The energy of ionizing radiation deposited per unit of body weight, corrected for "biological effectiveness" (i.e., the extent to which the particular exposure route leads to biological damage) is measured in sieverts (Sv).

Different exposure routes have different physiological impacts, and the biological effectiveness for inhalation and ingestion exposures to Pu-239 (in the form of PuO_2) is 1.2×10^{-4} Sv/Bq for inhalation and 9.7×10^{-7} Sv/Bq for ingestion.

The excess fatal cancer risk (equivalent to lifetime cancer risk used in earlier chapters—note that different terms can have the same meaning) due to alpha-radiation exposure is about 0.1/person-Sv. This means that a population exposure (i.e., summed over individual doses) of 10 Sv is likely to cause at least one fatal tumor. Radiation sickness from exposure of 10 Sv over a period of days will also almost certainly be lethal.

a. Suppose that 10 mCi of Pu-239 is the total inhaled by everyone in the world over some period of time. What will be the total dose? The per person dose?
b. What is the probability of contracting cancer from this source for each individual? How many total fatal cancer cases would you expect worldwide due to this exposure?
c. Suppose that 6 kg of Pu-239 (in the form of PuO_2 suspended as small particles) is dissolved in a reservoir of capacity $\sim 10^7$ m^3 serving a population of one million. What is the concentration of Pu-239 in this reservoir (assume even and rapid mixing, and that the threat is recognized after two days of consumption)?
d. What are the individual and population doses from this drinking water?
e. How many fatal cancers would you expect to be induced from this 6 kg of Pu-239 during the two days of exposure?

Solution 8-1

Solution 8-1a

The population dose is

$$10 \text{ mCi}_{Pu239} \times \frac{3.7 \times 10^{10} \text{ Bq}}{\text{Ci}} \times \frac{1.2 \times 10^{-4} \text{ Sv}}{\text{Bq}} = 44 \text{ Sv.}$$

The individual dose is

$$(44 \text{ Sv})/(6 \times 10^9 \text{ people}) = 7.3 \times 10^{-9} \text{ Sv/person.}$$

Solution 8-1b

$$\text{Population risk} = (44 \text{ Sv})(0.1 \text{ fatalities/Sv/person})$$
$$= 4.4 \text{ fatal cancers.}$$
$$\text{Individual risk} = (7.3 \times 10^{-9} \text{ Sv}) \times (0.1 \text{ fatalities/Sv})$$
$$= 7.3 \times 10^{-10} \text{ risk of fatal cancer.}$$

Note, however, that effects at such low exposures are highly uncertain. The Health Physics Society, for example, takes the position that at the exposure level in this section "risks of health effects are either too small to be observed or are nonexistent," and that only qualitative, and not quantitative, risk characterization is appropriate (Mossman et al. 1996).

Solution 8-1c

$$\text{Concentration} = \left(\frac{6 \text{ kg}_{\text{Pu 239}}}{10^7 \text{ m}^3 \text{ water}} \right) \left(\frac{1000\text{g}}{\text{kg}} \right) \left(\frac{6.2 \times 10^{-2} \text{Ci}}{\text{g}_{\text{Pu 239}}} \right)$$
$$\times \left(\frac{3.7 \times 10^{10} \text{Bq}}{\text{Ci}} \right) \left(\frac{\text{m}^3}{1000 \text{ liters}} \right)$$
$$= 1400 \text{ Bq/liter.}$$

Solution 8-1d

Assume that each person ingests about two liters of this water per day. Then

individual dose
$$= \left(\frac{2 \text{ liters}}{\text{person} \cdot \text{day}} \right) \left(\frac{1400 \text{ Bq}}{\text{liter}} \right) \left(\frac{9.7 \times 10^{-7} \text{ Sv}}{\text{Bq}} \right) (2 \text{ days})$$
$$= 0.005 \text{ Sv/person,}$$
population dose $= (0.005 \text{ Sv/person}) \times (10^6 \text{ people})$
$$= 5 \times 10^3 \text{ Sv.}$$

Solution 8-1e

$$\text{Number of fatalities} = (5 \times 10^3 \text{ Sv}) \times (0.1 \text{ fatalities}/\text{Sv})$$
$$= 500 \text{ cancer fatalities}.$$

Problem 8-A. Cassini Spacecraft Reentry Risk

This problem was suggested by Frank von Hippel.

Recently there was great controversy over the October 1997 launch of the $3.4 billion Cassini spacecraft on a journey to Saturn because it carries 30 kg of highly radioactive plutonium-238 (half-life of 88 years). This power supply is needed to provide electricity in a region where the solar intensity is only one percent of what it is near the Earth. The launch went without mishap. However, in August 1999, traveling at 12 miles/second after two swings-by of Venus, Cassini will pass the Earth at an altitude of 300 miles in a maneuver designed to further increase its speed for the trip to Saturn. The Pasadena-based Jet Propulsion Laboratory (JPL) manages the Cassini mission. The official position of NASA scientists at JPL is that the chance of an accidental reentry is below one in a million. That estimate is controversial, however.

If reentry were to occur, NASA estimated in its *Final Environmental Impact Statement* on the mission that 20–66% of the Pu-238 would be vaporized in the stratosphere (NASA 1995). This is approximately equal to the plutonium in the fallout from all atmospheric atomic bomb tests to date. If all of it were inhaled, it could induce cancer in the whole population of the earth. However, most of the dose we would receive would be from inhalation as the Pu-238 "falls out" of the atmosphere over a period of years at a very slow velocity—an estimated 2 cm/sec (Fetter and von Hippel 1990). Fortunately, however, only a small fraction would be inhaled.

a. What would be the transit time through a meter-thick layer of atmosphere of an average plutonium atom descending from the stratosphere?

b. During that time, what would be your chance of breathing in that atom? (If you breathe some from other layers, the plutonium atom will have passed through those layers also, so the result would be the same. An adult inhales about 20 m³ of air per day.)

c. What would the chance be that one of the six billion people on earth breathes in the atom?

d. How many grams of Pu-238 would be inhaled by the world's population?

e. How many cancers would be predicted to result? Assume 0.7 to 2.6×10^{-6} cancer deaths per gram Pu-238 inhaled (Fetter and von Hippel 1991).

Event Trees and Fault Trees

In the mid- to late 1800s, as steamboats became increasingly common in U.S. waters, so did the problem of boilers that could not withstand high temperature and pressures. These constituted both significant costs to the steamboat owners and hazards to passengers and operators. The question of whether to regulate the risk was characterized by the same arguments of disputed data, economic efficiency, and safety preferences that can be found in any risk regulation debate today, and eventually led to welding and materials standards.

In this area and others, such as mining and mass production, the increasing complexity of the industrial revolution forced improved approaches to systems analysis. One useful method is to isolate and trace system components, to see where failures might occur and identify weak links. Another is to look at possible failures, and back into what might cause those failures to occur. Both of these tools are common in both engineering and regulatory schemes.

Fault-tree analysis is a standard engineering approach for evaluating systems. It can also be a very useful tool for evaluating a range of nonengineering systems. Problems 8-2, 8-B, and 8-C explore increasingly complicated systems to illustrate the strengths and potential shortcomings that must be kept in mind when using fault-tree analysis. One benefit of this type of analysis is that it translates directly to a computer simulation of a risk-warning system. The associated drawback is that automated systems may be assumed to be "fail-safe" and as a consequence improperly monitored.

Two tools that can be useful for this type of analysis are schematic event trees and fault trees. Event trees look at a system and identify all possible outcomes. Fault trees identify the sequence of events that leads up to a single "fault" condition. For example, if you are filling a balloon with helium, there may be two possible outcomes: a full balloon and a popped balloon. A popped balloon might arise from two different conditions, that you overfilled it or that there was a flaw in the balloon. Figure 8-1a is an event tree for blowing up the balloon, and figure 8-1b is a fault tree showing events leading to an exploded balloon. (Note also that someone might pop the balloon!)

Also shown in figures 8-1a and b are probabilities associated with the possible outcomes. In this case, the probability of a flawed balloon is 1% and that of an overfilled balloon is 5%. In the fault tree, note the symbols. The rectangle represents an intermediate event, or a fault that results from a combination of prior events. Circles represent "basic events," those for which no further information is needed to understand the system, or given events. The bullet-like symbol with a concave bottom represents an "or" gate, meaning that a failure will occur with either overfilling or a flawed balloon, while the closed upside-down "U" represents an "and" gate, meaning that a balloon will only be overinflated if both the balloon is not flawed (since you won't have the opportunity to overfill a flawed balloon) and you put too much pressure in it. The "or" gate indicates that the sum of the probabilities of events leading to the final event is the probability of the final event, while the "and" gate indicates that the product of the probabilities is needed. (The symbols are those adopted by the U.S. Nuclear Regulatory Commission 1986.)

Problem 8-2. Simple Pressure Relief System

Suppose that you need to evaluate the probability that a pressure-tank relief system will work in case of overpressure; if the relief system fails, a weld will burst, destroying the tank. The tank has a primary pressure relief valve (PRV), which is backed up by a secondary system. You have good historical and technical information on both: the primary system (PRV 1) is expected to open successfully 99.9% of the time and the secondary system (PRV 2) 95% of the times it is needed.

   a. Draw an event tree describing this system.
   b. Use this tree to calculate the probability of a tank failure given overpressure.

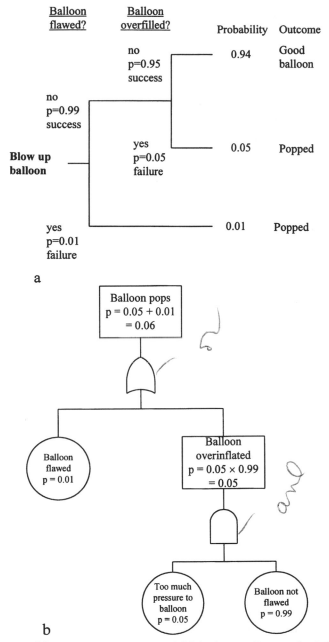

Figure 8-1. (a) Event-tree diagram and (b) fault-tree diagram for inflating balloon.

Now, think about how this pressure tank fits into a larger system. Assume that the tank holds high-pressure steam, a hydrocarbon mixture, and a catalyst, and that a single operation of the tank takes ten hours. The pressure is regulated by a computer-operated control circuit, designed to keep the system between specified minimum and maximum values. If the computer fails, a light comes on in a control room to signal an operator to regulate pressure manually. Assume that the steam supply exceeds normal pressure 0.1% of the time, that the computer controls overpressures 98.5% of the time, and that the operator's time is allotted such that she is available to see and react to the light 75% of the time. Figure 8-2 is a "piping and instrumentation diagram" (P & ID) describing this system.

c. Draw an event tree for this system, and a fault tree for unsafe operations. Calculate the probability that the backup valve will be needed given excess pressure from the steam supply, and the overall probability that the tank will fail in a single operation. What is the probability of 100 consecutive safe operations?

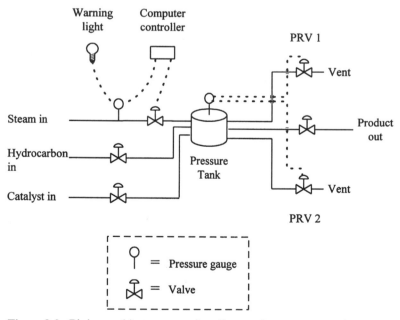

Figure 8-2. Piping and instrumentation diagram for pressure tank system.

Solution 8-2

Solution 8-2a

Figure 8-3 shows the event tree for this system.

Solution 8-2b

This calculation is shown in figure 8-3. Since a system failure occurs only when both the primary and secondary valves fail, the total probability of failure is the product of the two failure probabilities, or

$$P(\text{failure}) = P(\text{primary failure}) \times P(\text{secondary failure})$$
$$= 0.001 \times 0.05$$
$$= 0.00005$$

Solution 8-2c

This system, shown in figure 8-4, is clearly more complex than the subsystem originally identified. However, while creating an event-tree diagram requires more time, it requires no fundamentally new methodology.

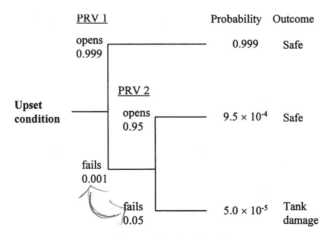

Figure 8-3. Event-tree diagram for single tank overpressure event.

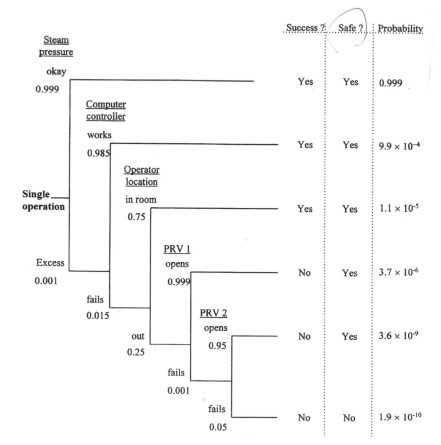

Figure 8-4. Event-tree diagram for problem 8-2c.

Figure 8-5 shows a fault tree for the events leading up to an unsafe operation. In the system we have presented here, fault trees and event trees can represent outcomes equally well. However, fault trees are frequently more useful for analyzing more complex engineering systems, since they can provide clearer pictorial presentations and can be used to identify critical events leading to a single type of failure. For example, in a system with ten possible outcomes (as opposed to the two-outcome system presented here), the final column in an event tree would become incomprehensible. Using fault trees would allow the analyst to consider each of the ten failure modes individually.

The probability that the backup is needed because of a pressure excess means that the computer controller has failed, the operator was

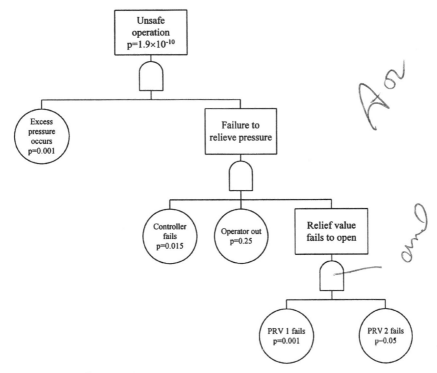

Figure 8-5. Fault-tree diagram for problem 8-2c.

not in the control room, and the primary valve has failed to open, so

P(PRV 2 needed | pressure excess)

   $= P$(computer fails and operator out and PRV 1 fails)

   $= P$(computer fails) $\times P$(operator out) $\times P$(PRV 1 fails)

   $= 0.015 \times 0.25 \times 0.001$

   $= 3.8 \times 10^{-6}$.

The overall probability of a failure for each operation is given in figure 8-3: 1.9×10^{-10}. The probability of 100 consecutive safe operations is equivalent to the probability of no unsafe operations. Since the probability of an unsafe operation is 1.9×10^{-10} each time,

$$P(100 \text{ consecutive safe operations})$$

$$= (1 - 1.9 \times 10^{-10})^{100}$$

$$= 0.999999981.$$

Problem 8-B. Additional Fault Tree

Construct a fault tree for unsuccessful operations.

Problem 8-C. Calculations from Fault Trees

What is the probability of a successful operation? Of a safe operation? Of 1000 consecutive successful operations? How many operations would you need to have a 5% risk of a failure?

Problem 8-3. Missing Components, Common-Mode Failures, and the Human Element

As systems become increasingly complex and interconnected, the potential to miss possible failure pathways increases. The problem is further complicated by the difficulties inherent in modeling human behavior as an input. Numerous important policy decisions have been informed by, and major catastrophes caused by, risk analyses with one or more of these shortcomings. Examples include the DC-3 case in chapter 2 (where a common-mode failure was assumed to involve two independent modes), the WASH-1400 report on nuclear reactor safety in chapter 1 (where important failure modes were left out entirely), and the Bophal, India accident (where training and practices were inconsistent with the design criteria). This problem revisits the system in problem 8-2 to explore some of these issues.

Evaluate and discuss how each of the following conditions affects the operation and overall safety of the system in figure 8-2:

   a. 40% of the primary safety system failures are because the pressure meter on the tank, which signals both the primary and backup valves, reads incorrectly. Assume that the 5% failure rate of the backup system was based on mechanical failure.
   b. An incorrect mixture of hydrocarbon and catalyst is three times as likely to cause the system to overpressurize than is high steam pressure.

c. A misprint in the operator's manual says to turn the steam valve counterclockwise in case of a warning light, when it should read clockwise.

Solution 8-3

Solution 8-3a

This is a classic common-mode failure problem. In this system, both the *primary* and the *secondary* systems receive data on system performance from the same pressure meter. In 40% of the failure cases, the cautiously redundant and seemingly well-planned backup systems will be useless because of the problem with the pressure meter.

Based on this new information, we can expect the backup valve to fail due to incorrect pressure reading on 40% of the occasions that it is needed, and, due to mechanical failure, 5% of the time that the pressure reading is correct. Consequently, the probability of a backup system failure is $(0.4 + 0.6 \times 0.05) = 0.43$ (and not 0.05), and the overall probability of a system failure is 1.6×10^{-9}, an order of magnitude greater than the originally calculated 1.9×10^{-10}.

Solution 8-3b

This is an example of a missing element. It affects the pressure in the tank only, while the gauge is on the steam line. Consequently, the high pressure is noted by neither the computer nor the operator. This means that there is a 0.4% chance of high tank pressure with each operation. The probability of a steam pressure excess followed by failure of both the computer and operator to respond is comparably negligible at 3.8×10^{-4}. Consequently, the probability of a failure is

$P(\text{failure})$

$\qquad = P(\text{excess pressure}) \times P(\text{PRV 1 fails}) \times P(\text{PRV 2 fails})$

$\qquad = 0.004 \times 0.001 \times 0.05$

$\qquad = 2 \times 10^{-7},$

which is three orders of magnitude greater than the probability predicted in problem 8-2c.

Solution 8-3c

This condition effectively removes the operator as a safety system component. Figure 8-6 shows the event tree modified for this situation. Removing the operator from the system, the probability of an unsuccessful operation is an order of magnitude greater, and the probability of an unsafe operation is four times as great.

Problem 8-D. Additional Fault Trees

Create a fault tree for unsuccessful and one for unsafe operations based on figure 8-6.

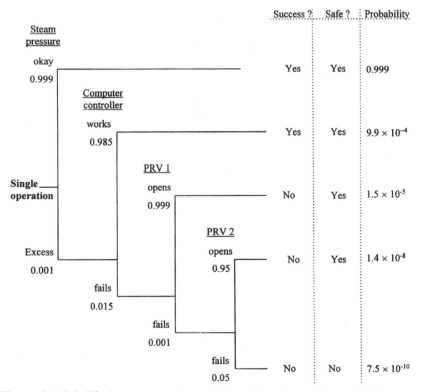

Figure 8-6. Modified event tree for problem 8-3c. Comparison with figure 8-4 indicates that the probability of an unsuccessful operation is about four times as great in the absence of the operator.

Problem 8-E. Identifying Problems

   a. Suggest some other potential common failure modes in figure 8-2. Evaluate and discuss their impact on the risk analysis.

   b. Repeat (a) for missing components.

   c. Draw a fault tree describing problem 8-3a as a missing component (rather than a common-mode failure) situation.

Problem 8-F. Anticipating the Unknown

Suggest five fundamentally different ways that an operator could take an incorrect or inadequate action given a valid warning light. How might one anticipate and prepare for each?

Problem 8-G. Oleum and the "Clever-Proofing" Problem

It is not unusual to hear about safety systems that are "foolproof." This is so common, in fact, that we often forget that this means that the system is designed to be safe even in the case of an inept operator. One of the authors (DMH), however, has suggested the term "clever-proof" for a less well publicized problem: the operator who intentionally subverts a safety system when he or she believes it will make the overall operation go more smoothly. For example, a remote activation button was installed at a metal stamping machine to insure that the operator was clear of the machine before each operation. An operator who used a broom handle to press the button—and speed up the job—found that a broken hand slows things down considerably.

   In 1994, following an event where several untrained workers were told to empty a railroad tanker car of oleum (supersaturated sulfuric acid) and ended up sending a plume of concentrated sulfuric acid fog into a populated area, local authorities decided to significantly increase the safety at companies that handled oleum. At one facility (with no record of oleum mismanagement), a new loading system was installed in which a tanker truck driver had to hook up all of the necessary piping and then go into a shack and sit on a pressure-sensitive chair while observing the loading operation. If the operator were to leave the shack, the chair would send a signal shutting down the system.

Suggest at least two ways in which the operator could disable these safety systems, and reasons why she or he might choose to do so.

Problem 8-4. Coal-Burning Power Plant Emissions

Consider a coal-burning power plant rated at 1,000 electrical megawatts, which generates 18,000 megawatt-hours of electricity per 24-hour day—corresponding to a "load factor" (actual output divided by potential output at rated capacity) of

$$\left[\frac{\dfrac{18,000 \text{ MWh}}{\text{day}}}{\dfrac{24,000 \text{ MWh}}{\text{day}}} \right] = 0.75.$$

Suppose this plant has a heat rate (energy needed to generate a kilowatt-hour of electricity) of 10 MJ/kWh, burns coal containing 30 MJ/kg and 2% sulfur by weight (0.02 kg S/kg coal), and uses sulfur-control technology that meets the EPA's New Source Performance Standard by removing 90% of the sulfur oxides from the stack gases.

a. Calculate the annual emissions of sulfur dioxide from this power plant in tons.
b. Estimate the annual number of excess deaths and the person-years of lost life expectancy attributable to the operation of this power plant for two combinations of assumptions about plant site and dose-response relation:

i. 10 person-μg/m^3 exposure per annual ton sulfur and the National Academy of Sciences (NAS) dose-response estimate of 2 excess deaths and 0.2 years lost life expectancy per 10^6 person-μg/m^3; and
ii. 200 person-μg/m^3 per annual ton sulfur, and the Brookhaven National Laboratory (BNL) dose-response estimate of 37 excess

deaths and 370 years lost life expectancy per 10^6 person-$\mu g/m^3$.

c. Suppose this harm is inflicted on the same 0.75 million Americans whose electricity is supplied by the plant. Estimate the annual death rate in this population from other causes. (You can look it up if you want, but a simple stock flow–residence time argument using the life expectancy at birth as the residence time will produce a decent estimate.) Compare this "natural" death rate to the incremental death rates attributable to the power plant for the two cases considered in part b. If cancer causes 20% of the total deaths in the population and each cancer death costs 15 years of life expectancy, how does the total loss of life expectancy attributable to the coal plant compare to that from all cancer? Comment on what seems to you to be most interesting (or uninteresting) about these results.

Solution 8-4

Solution 8-4a

The amount of coal used is

$$\left(\frac{18000 \text{ MWh}}{\text{day}} \right)\left(\frac{1000 \text{ kWh}}{\text{MWh}} \right)\left(\frac{10 \text{ MJ}}{\text{MWh}} \right)\left(\frac{1 \text{ kg coal}}{10 \text{ MJ}} \right)$$

$$= 6 \times 10^6 \text{ kg coal.}$$

The amount of SO_2 produced is given by

$$(6 \times 10^6 \text{ kg coal})\left(\frac{0.02 \text{ kg S}}{\text{kg coal}} \right)\left(\frac{2 \text{ kg SO}_2}{\text{kg S}} \right)\left(\frac{1 \text{ ton SO}_2}{1000 \text{ kg SO}_2} \right)$$

$$\times \left(\frac{0.1 \text{ ton SO}_2 \text{ emitted}}{\text{ton SO}_2 \text{ produced}} \right)\left(\frac{365 \text{ days}}{\text{year}} \right)$$

$$= 8{,}800 \text{ tons SO}_2 \text{ emitted each year.}$$

Solution 8-4b(i)

The excess deaths are given by

$$(8800 \text{ tons SO}_2)\left(\frac{0.5 \text{ tons S}}{\text{tons SO}_2}\right)\left(\frac{10 \text{ person } \mu g/m^3}{\text{annual tons S}}\right)$$

$$\times \left(\frac{2 \times 10^{-6} \text{ deaths}}{\text{person-}\mu g/m^3}\right)$$

$$= 0.09 \text{ deaths},$$

and the lost life expectancy is

$$(0.09 \text{ deaths})\left(\frac{0.1 \text{ year}}{\text{death}}\right) = 0.009 \text{ year}.$$

Solution 8-4b(ii)

In this case, excess deaths are

$$(8800 \text{ tons SO}_2)\left(\frac{0.05 \text{ tons S}}{\text{tons SO}_2}\right)\left(\frac{200 \text{ person-}\mu g/m^3}{\text{annual tons S}}\right)$$

$$\times \left(\frac{37 \times 10^{-6} \text{ deaths}}{\text{person-}\mu g/m^3}\right)$$

$$= 33 \text{ deaths},$$

and lost life expectancy is

$$(33 \text{ deaths})\left(\frac{10 \text{ yr}}{\text{death}}\right) = 330 \text{ years}.$$

Solution 8-4c

$$\text{Flow} = \frac{\text{stock}}{\text{residence time}}.$$

Assume that residence time = life expectancy at birth, life expectancy = 75 years, and stock = people served by plant = 750,000 people. Then

$$\text{flow} = \frac{750,000 \text{ people}}{75 \text{ year}} = 10,000 \text{ people/year}.$$

If cancer causes 20% of the total deaths in the population and the associated loss of life expectancy is 15 years/death, the total annual loss of life expectancy due to cancer is

$$(10000 \text{ deaths/year})\left(\frac{0.2 \text{ cancer death}}{\text{total deaths}}\right)\left(\frac{15 \text{ years lost}}{\text{cancer death}}\right)$$

$$= 30,000 \text{ years}.$$

The loss of life expectancy attributable to the coal plant is six orders of magnitude less than that due to cancer in the population served by the plant using a best-site/NAS-dose-response calculation but only two orders of magnitude less using a worst-site/BNL-dose-response calculation.

What is most interesting about these results is that the coal plant is found to be either negligible or significant depending entirely on assumptions about site and dose-response relation.

What might be called uninteresting is that the comparison between coal plant consequences and total cancer consequences is not very meaningful. More interesting is finding an explanation for the other cancers, or a comparison of the consequences of alternative ways to get the same benefit—electrical services.

Problem 8-5. Commercial Nuclear Power Safety: An Empirical Analysis

The United States had about 10 large (approximately 1 GWe) commercial nuclear power reactors on line in 1970, about 60 in 1980, and about 110 in 1990. Since 1990, the number has plateaued. The Three Mile Island accident, the most serious nuclear reactor accident in U.S. history, occurred in March 1979. However, it did not release a significant amount of radioactivity.

a. About how many reactor-years of nuclear power experience did the United States have at the beginning of 1979?
b. Roughly how many U.S. reactor-years had accrued by the end of 1997 without a major release of radioactivity?
c. Suppose that a serious accident occurred tomorrow and the released radioactivity ultimately caused an estimated 10,000 cancer deaths (a Chernobyl-scale accident). Based on that singe event, what would you estimate for the cancer deaths per nuclear-reactor year?
d. What yardsticks would you use if you wanted to decide whether or not this would be an acceptable level of risk?

Solution 8-5

Solution 8-5a

Approximating the data prior to 1990 with a straight line, the number of U.S. nuclear power plants (N) in years (Y) prior to 1990 was

$$N = 5 \cdot (Y - 1968),$$

and for after 1990,

$$N = 110.$$

Based on this, the number of reactor-years prior to 1979 (the area of a triangle with a base from 1968 through 1978 and a height of N_{1979}) would be

experience
$$= 0.5 \cdot [(1978 - 1968) \text{ years}] \times [5 \cdot (1978 - 1968) \text{ reactors}]$$
$$= 250 \text{ reactor-years}.$$

Solution 8-5b

The number of reactor years through 1990 was

experience
$$= 0.5 \cdot [(1990 - 1068) \text{ years}] \times [5 \cdot (1990 - 1968) \text{ reactors}]$$
$$= 1210 \text{ reactor-years}.$$

The number of reactor-years since 1990 is simply 110 times the number of years, so the total experience through 1997 was

$$\text{experience} = [1210 + 100 \times (1997 - 1990)] \text{ reactor-years}$$
$$= 1,980 \text{ reactor-years}.$$

Solution 8-5c

$$(10,000 \text{ cancer deaths}/1980 \text{ reactor-years})$$
$$= 5 \text{ cancer deaths per reactor-year}.$$

Solution 8-5d

One possibility would be to compare these numbers to the estimated deaths per year from an alternative source of power. The main alternatives to nuclear are fossil fuels, which currently generate about 6% of the fine particulate pollution to which the U.S. EPA attributes ten thousand deaths per year and one-third of the nitrogen oxide precursors to tropospheric ozone which (again according to the U.S. EPA) causes 1.5 million respiratory injuries per year (*New York Times*, 28 November 1996, p. 1). These estimates are very rough, however—like the upper bound estimate suggested in (c) above. Renewable energy sources will also have some external costs, but they are difficult to estimate in the absence of concrete design systems.

On a basis of release-related deaths alone, it is difficult to reject nuclear power. The next problem takes a broader look, and considers some intractable problems associated with the disposal of high-level nuclear wastes.

Problem 8-H. The Risk from Nuclear Accidents

This problem was contributed by Tony Nero.

Consider figures 8-7 and 8-8, which are fatality estimates from the WASH-1400 report discussed in chapter 1 (U.S. Nuclear Regulatory

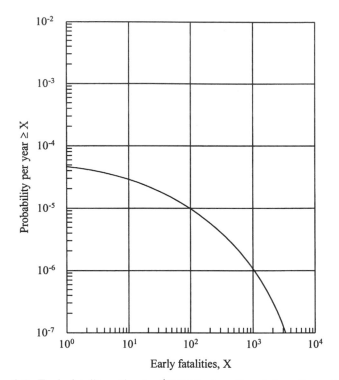

Figure 8-7. Early fatality estimates (U.S. Nuclear Regulatory Commission 1975). "This figure shows the probability that in a given year nuclear plant accidents will cause a minimum of the specified number of early fatalities, i.e., deaths from acute radiation exposure. The probability given assumes that 100 plants, of the type studied by WASH-1400, are operating. Note for example that the probability of one death occurring in a given year was found to be less than one in ten thousand (for 100 reactors)" (Nero 1979).

Commission 1975). These figures give the probability per year that accidents will cause fatalities or fatalities per year greater than X plotted versus the number X. The popular materials describing the results left the impression that most of the total risk of fatalities was from accidents that caused relatively few fatalities. However, whether this is true is not clear from the figures, which represent probabilities, not risks, and even then in a cumulative manner (i.e., the probability of fatalities *above* the specified level).

Using the figures, extract the probability (not cumulative) versus small consequence intervals (in terms of small ranges of fatalities, X) so

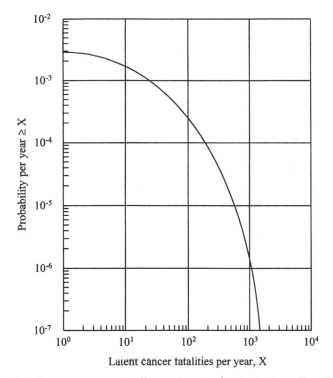

Figure 8-8. Latent cancer fatality estimates (U.S. Nuclear Regulatory Commission 1975). "This figure shows the probability that in a given year nuclear plant accidents at any of 100 reactors will cause a minimum of a certain number of cancer fatalities per year during a 30 year latency period after the accidents. To get the total number of cancer fatalities caused versus the annual probability of this minimum, the scale at the bottom should be multiplied by 30" (Nero 1979).

that you can then estimate risk (as a product of probability and consequences) associated with each consequence interval.

a. For each of the figures, present your estimates in tabular form.

b. For each type of consequence (early, latent), estimate from your table the magnitude of accident (in terms of consequences) above which one-quarter, one-half, and three-quarters of the total risk occurs.

c. If you integrate the risk for the two types of fatalities, do you get totals similar to those given in table 8-1?

Table 8-1 WASH-1400 Annual Risk Estimates for One Hundred
Nuclear Power Plants

Consequence	Societal	Individual
Early fatalities	3×10^{-3}	2×10^{-10}
Latent cancer fatalities	7×10^{-1}/year	3×10^{-10}/year

Source: Nero 1979.

Problem 8-6. Long-Term Risks from High-Level Nuclear Waste: A Case of Extreme Uncertainty

High-level nuclear waste (HLNW) remains a potential human and ecological health hazard for about 10^4 years after it is generated. Consequently, as we decide how to manage this waste, which is produced in military weapons manufacture and civilian nuclear power facilities, we must consider its long-term fate. Yucca Mountain, Nevada (about one hundred miles north of Las Vegas), is currently the only site being considered by the Department of Energy (DOE) as the permanent storage facility for an eventual 70,000 tons of United States' HLNW.

Two hotly contested and highly uncertain issues in the licensing debate are the prospects for preventing human intervention (intentional and unintentional) and the geologic stability of the area. This problem evaluates some approaches for dealing with these two issues.

    a. Evaluate the assertion that disposing of radioactive wastes is a "trivial technical problem" (Cohen 1990, as cited in Flynn et al. 1997). John Cantlon (1996) of the Nuclear Waste Technical Review Board points out that "there is no scientific basis for predicting the probability of inadvertent human intrusion over the long times of interest for a Yucca Mountain repository." He goes on to argue that "intrusion analyses should not be required and should not be used during licensing to determine the acceptability of the candidate repository." In contrast, Flynn et al. are unconvinced that "the current inability to assess a risk is reason for eliminating it from consideration in the licensing process."

b. Why is human intrusion a concern, and why is there such great uncertainty about it? What measures might be taken to prevent it? Is it necessary to do so?

c. Evaluate the argument that if information on a risk is neither available nor forthcoming, it should not be considered.

This section is based on Kerr (1996). One concern about Yucca Mountain is that, even if the facility will be able to contain the HLNW for ten thousand years in the absence of geologic disruption, a major volcanic eruption could release substantial amounts. Within a forty kilometer radius of the proposed facility, there are about a dozen dormant volcanoes, and while there is considerable disagreement on how active these volcanoes are, it is possible that one has been active as recently as twenty thousand years ago.

Because of the controversy and expert disagreement surrounding potential volcanic activity, the DOE assembled ten prominent earth scientists and used a method called "expert elicitation" to mathematically combine their evaluations of expected values and distributions. The results of this expert elicitation are shown in figure 8-9.

Kerr describes the expert elicitation process as follows:

First, Coppersmith [the project leader] and crew chose the 10 panel members on the basis of expertise and institutional affiliation—plus "strong communication and interpersonal skills, flexibility and impartiality." Workshops and field trips with outside experts followed. Then interviewers spent two days

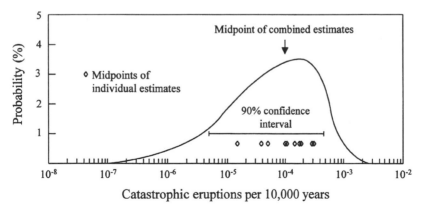

Figure 8-9. Probability of magma intrusion into the proposed Yucca Mountain depository as elicited from experts by Geomatrix Consultants, Inc.

with each panel member, extracting the expert's best estimate for values such as the location and frequency of expected eruptions, and the accompanying uncertainties for each parameter. Those parameters were then plugged into a chain of calculations leading to the probability that a magma conduit would cut through the repository. The 10 experts' resulting probabilities—and associated uncertainties—were eventually combined into the aggregate probability.

d. Evaluate the interpretation that "the panel of ten earth scientists concluded that the probability of a volcano erupting through the repository during the next ten thousand years is about 1 in 10,000."

e. Robert Miller, the governor of Nevada, argued that the evidence for recent volcanism was sufficient to disqualify Yucca Mountain as a potential site. Should potential volcanic activity be a "cutoff" criterion, or should it be considered as one input into an overall analysis?

f. Discuss possible advantages and shortcomings of an expert elicitation approach.

Solution 8-6

Solution 8-6a

The long time horizon and the uncertainties associated with such an amount of time suggest that disposing of radioactive wastes is considerably more than a technical issue. Instead, it is an issue with which our society continues to grapple, as well as one in which "analysis" is often proxy, rather than support, for preferences. Consequently, knowing what the uncertainties and preferences are in the issue help clarify where better analysis is useful.

Solution 8-6b

Human intrusion is a concern because it could mean anything from accidental exposure by a few accidental visitors to intentional and widespread exposure. On the other hand, since it is only in the past sixty years that people have concentrated large amounts of highly radioactive materials, it is possible that intervention in the future could be in the form of widespread beneficial use. It is the uncertainty about what society and the state of technology will be like over the next 10,000 years that generates uncertainty about potential future intrusion.

It is possible to design the facility to avoid intrusion—for example, by maintaining guards, sealing off compartments as they are filled, posting a variety of warnings. However, any conceivable design could also conceivably be overcome. Certainly, it seems unlikely that the government will be stable for 10,000 years, so any design that requires ongoing action is not a real option. Similarly, warning systems might not be comprehensible in 5000 or even 500 years, and while they could divert accidental intrusion, they could also attract intentional intrusion.

Whether we should try to prevent intrusion over such a long time period is an issue that no amount of analysis can resolve. Individual preferences will dominate this discussion, with some concluding that future societies will be better able to deal with this waste than can we, while others conclude that we should take responsibility for our own wastes and not leave it to future generations.

Solution 8-6c

This is a problematic assertion. Even in a case such as this, where it is unlikely that uncertainty can be reduced within the time scale of the decision, ignoring a potential problem is unwise. Instead, the existence of such uncertainties might help differentiate among options. In this example, knowing that there is an unsolvable problem might suggest that we expand the decision horizon to include enough time to solve it.

Solution 8-6d

The distribution, as well as the way in which it was generated, suggests that using the central tendency to the exclusion of a more thorough interpretation of the available information is problematic. Kerr goes on to say that "the 90% confidence interval runs from 5 chances in 1 million to 5 chances in 10,000." This is considerably more informative, since it suggests that, based on this research, there is a 5% probability that there will be multiple magma intrusions. Without such distributional information, we have no basis for believing that the 90% confidence interval is more narrow or includes one in one hundred! Nonetheless, it is possible that decision makers will stop at the central tendency estimate, and ignore the distribution.

Solution 8-6e

Governor Miller's assertion highlights the difficult nature of low-probability, high-consequence, high-uncertainty issues. It is impossible to disagree with him on any "technical" terms, but many would argue that his decision to allow one criterion to dominate ignores alternatives and other possible outcomes of this solution. It also indicates how analysis can be used in the policy process, since Miller wants to use a single piece of analysis to reject the Yucca Mountain site, while we saw above that Cantlon (a proponent of the site) argued that we should reject analysis that does not support his view.

Solution 8-6f

Since this issue was rife with dissent and uncertainty, expert elicitation has the potential to produce a general idea of the probabilities and generate discussion among experts that can lead to more complete understanding, both of the information and of additional information needs. There are possible concerns, however. These include not only the issue of who is chosen to be on the panel, but also other issues:

- What form do the interactions take? "Groupthink," where the goal of arriving at consensus undermines individuals' willingness to dissent (Kleindorfer et al. 1993) can distort consensus-seeking processes.
- Morgan and Henrion (1990) document a number of cases in which experts underestimate the uncertainty of their own estimates. There have been some attempts to identify ways to include this "expert uncertainty" in estimates (e.g., Shlyakhter 1994), but a healthy dose of skepticism remains a valuable tool in the face of large and important uncertainty.
- Finally, the mathematics involved in averaging expert opinions that are already averages from a number of models can artificially

narrow the combined distributions. (Geologist Eugene Smith and mathematician Chih-Hsiang Ho, both of the University of Nevada, Las Vegas, as cited in Kerr 1996).

Problem 8-7. Military Fighter Aircraft

In September of 1997, U.S. Secretary of Defense Cohen ordered a day-long hiatus of nonessential military flights following a one-week period in which there were six unrelated military aircraft accidents. Table 8-2 contains some statistics on military aircraft flight safety. The Pentagon argued that (a) overall flight safety was improving each year; (b) the series of six accidents in seven days was a statistical anomaly; but (c) they would undertake a major review of flight safety anyway. This problem looks at each of these assertions in turn.

a. Is overall flight safety improving?

   i. Assume that there were about 50,000 total military flights each year. Is the reduction in number of major accidents statistically significant?

   ii. The military uses number of major accidents per 100,000 flight hours as a yardstick for flight safety. Calculate this number for each year. Is there a statistically significant change in this rate from 1994 to 1997?

   iii. Why might accidents per unit flight time be preferable to total accidents per year or total accidents per flight (table 8-3)?

b. Were the six accidents in one week statistically anomalous?

Table 8-2 U.S. Military Aircraft Accidents, by Year

Year	Major accidents
1997	54
1996	67
1995	69
1994	86

Source: Jane's Defense Weekly 9/24/97.

Table 8-3 U.S. Military Aircraft Flight Hours, by Year

Year	Flight hours
1997	3.60×10^6
1996	4.47×10^6
1995	$4.44 \times 10^{6\,a}$
1994	5.31×10^6

Source: Jane's Defense Weekly 9/24/97.
[a] Estimated.

c. If the military believes that overall flight safety is improving, and the six accidents in one week were statistically anomalous, why did Secretary Cohen use them as a signal to undertake a major review?

Solution 8-7

Solution 8-7a(i)

A starting point would be to consider the extreme years (1994 and 1997); if the difference between these is insignificant, then the difference among the others must be insignificant as well. To test the hypothesis that 1994 is the same as 1997, use the familiar Z-score, with $\bar{x} = 54$, $p = 86/50{,}000 = 1.7 \times 10^{-3}$, and $n = 50{,}000$. Recalling equation 3-7c,

$$Z = \frac{\bar{x} - np}{\sqrt{np(1 - p)}}$$

$$= \frac{54 - 86}{\sqrt{50000 \cdot 1.7 \times 10^{-3}(1 - 1.7 \times 10^{-3})}}$$

$$= -3.5,$$

which is significant at the 95% level, so we can reject the hypothesis that the accident rates are the same at the 95% confidence level.

Solution 8-7a(ii)

First, we can combine tables 8-2 and 8-3 to find accidents per hundred thousand flight hours (table 8-4).

Clearly, there is no change in flight safety from 1996 to 1997 according to this measure. The difference from 1994 to 1997, calculated as above, is not significant at the 95% level ($p = 0.26$).

Solution 8-7a(iii)

Choosing the unit of analysis is an important (and generally value-laden) decision. However, there are clearly some strong reasons to prefer something other than absolute numbers. Consider, for example, the case of a year in which there were several orders of magnitude fewer flights. If the number of accidents did not drop, one would suspect that safety practices were becoming lax.

Given the data on total flight hours, it appears that the assumption that year to year total number of flights was constant was probably not valid and consequently misleading. If most accidents occur during takeoff or landing, total number of flights might be more useful than the flight-hour number.

Solution 8-7b

This type of analysis can be done with another statistical technique, known as the Poisson method, which is an approximation of the binomial for rare events. As opposed to most of statistics, which was

Table 8-4 U.S. Military Aircraft Flight and Accident Data

Year	Major accidents	Flight hours	Accidents / 10^5 flight hours
1997	54	3.60×10^6	1.50
1996	67	4.47×10^6	1.50
1995	69	4.44×10^{6} [a]	1.55
1994	86	5.31×10^6	1.62

[a] Estimated.

commissioned by wealthy gamblers, the application that brought the Poisson method attention was very similar to this problem. In the late 1800s the Prussian military was concerned whether a rash of cavalry officers kicked to death by their horses represented a real change in behavior or a statistical fluke; the Poisson calculation suggested that it was the latter.

The Poisson formulation is

$$p(x) = \frac{e^{-\lambda} \cdot \lambda^x}{x!},$$

where λ is the expected frequency, x is the discrete value for which you are testing, and $p(x)$ is the probability of that specific outcome. Since there is normally about one accident per week, $\lambda = 1$, and you're testing for the observed number of accidents, $x = 6$. Since

$$p(6) = \frac{e^{-1} \cdot 1^6}{6!} = 0.0005,$$

one would expect six accidents in a single week once out of every two thousand weeks, or one week in every 40 years.

From this perspective, then, the six accidents in a week are statistically significant! (Note that you would get similar results using the Z-score method: $p < 0.0001$.) Again, however, unit of analysis and independence become important issues. Which is more important, the no-change finding for the year or the significant-change finding for the week in question? If these accidents were truly independent, then even the "statistically significant" difference has no practical meaning.

Solution 8-7c

While both assertions (overall improvement, irrelevance of short-term numerical increase) were incorrect in some sense, the reaction was probably reasonable. This leads to a very important issue often neglected in risk analysis: what is the relevant baseline? If the existing baseline can lead to a statistically insignificant but nonetheless eye-opening blip, it may be that the baseline was unacceptable, and the observed cluster is a signal that some remedial action should be taken. In short, an assertion that a deviation from the norm is not important because it is not statistically significant implies that the underlying base rate was acceptable. This is more appropriately considered a priori.

Problem 8-I. Who Decides What's Important?

The media and the military both mentioned the decrease in the total number of accidents much more often than the "officially preferred" number of accidents per flight hour. Why do you think this is the case?

Problem 8-8. Risk of Domestic Airplane Flight

Table 8-5 summarizes accident data for major U.S. airlines, as of 1993. One of the major airlines that serve domestic routes (Southwest) had unblemished safety records—no deaths from accidents

  a. Assuming that each airline's safety record reflects its inherent safety, what are the chances of dying from flying on Southwest?

Table 8-5 Accident Data for Major U.S. Airlines

United States airlines	Death risk per flight [a]
U.S. Air (US)	1 in 2.5 million
Northwest (NW)	1 in 4 million
TWA (TW)	1 in 5 million
United (UA)	1 in 8 million
American (AA)	1 in 10 million
Delta (DL)	1 in 10 million
Continental (CO)	1 in 15 million
Southwest (WN)	0 (perfect record)
Industry average	1 in 7 million

[a] *Source: Newsweek* 1994.
This statistic is the probability that someone who randomly selected one of the airline's flights over the twenty-year study period would be killed en route.

300 • CHAPTER 8

b. The other extreme assumption would be the possibility that all airlines are identical, and the observed differences are entirely random. If this is true, what is the probability of dying from flying on Southwest? On U.S. Air?

c. Alternatively, it is possible that some inbetween value is the true probability for each airline—that is, there is both a random and a causal element in each airline's record. However, this cannot be evaluated from the data given in this problem. Use Bayesian analysis to evaluate the risks of flying U.S. Air. Be explicit about assumptions and uncertainty.

Solution 8-8

Solution 8-8a

All that can be said about flying Southwest is that there is an upper bound on the expected death risk. For example, making the assumption that Southwest does not have the smallest total number of flights, then one can assume that there is less than a 1 in 15 million risk associated with flying Southwest.

Solution 8-8b

If the differences are random, then the real risk on either airline is the industry average of 1 death per 7 million passenger flights.

Solution 8-8c

Recall from chapter 4 Bayes theorem (equation 4-4):

$$P(A_j \mid B) = \frac{P(B \mid A_j) \times P(A_j)}{\sum_{i=1}^{n} P(B \mid A_j) \times P(A_j)}.$$

In this case we condition the prior risk ($1{:}2.5 \times 10^6$ for U.S. Air) on "new" information, the industry average. Assuming that all airlines are about the same size (or at least that U.S. Air is of average size), the probability that any particular flight is a U.S. Air flight is $1/8$ (or 0.125), and the industry average excluding U.S. Air is still about 1 in 7 million [$7/(4 + 5 + 8 + 10 + 10 + 15) \times 10^6$]. Also, the prior probability that U.S. Air was the airline used given an accident was [$(1/2.5)/(1/2.5 + 1/4 + 1/5 + 1/8 + 1/10 + 1/10 + 1/15)$]) = 0.32. Then the posterior risk of an accident on U.S. Air is

P(U.S. Air | prior and new information)

$$= \frac{P(\text{accident} \mid \text{U.S. Air}) \times P(\text{U.S. Air})}{P(\text{accident} \mid \text{U.S. Air}) \times P(\text{U.S. Air}) + P(\text{accident} \mid \text{not U.S. Air}) \times P(\text{not U.S. Air})}$$

$$= \frac{(1/2.5) \times 10^6 \times (1/8)}{(1/2.5) \times 10^6 \times (1/8) + (1/7) \times 10^6 \times (7/8)}$$

$$= 0.29.$$

Note, however, that this outcome depends on how the prior and new information are defined and weighted. It might well be as reasonable just to say that the "real" probability that U.S. Air will be the airline given that there is an accident is somewhere between 0.125 and 0.32.

Problem 8-J. Verifast Airlines

Suppose that Verifast Airlines, a new company, opened, and on its first day of operation had two major accidents resulting in 159 deaths out of 1000 person-flights. Three detailed investigations, one by a consumer group, one by the airline, and one by the Federal Aviation Authority, all indicate no unusual safety problems. How would you determine the "risk" of flying on Verifast? Is it a 20% risk of death?

Problem 8-K. Risk of International Airplane Flight

Repeat problem 8-8 using the world flight data in table 8-6.

Table 8-6 Accident Data for Major International Airlines

International airlines	Death risk per flight
Aeroflot	1 in 200,000
Air India	1 in 200,000
Egypt Air	1 in 200,000
Malev (Hungary)	1 in 200,000
Varig	1 in 200,000
Aeromexico	1 in 300,000
Air Canada	1 in 1.5 million
Japan Airlines	1 in 2 million
British Airways	1 in 3 million
Lufthansa	1 in 5 million
El Al	1 in 20 million
Argentinas	0
Singapore	0
Industrialized countries	1 in 3.5 million
Developing nations	1 in 500,000

References

Cantlon, J. (1996). *Nuclear Waste Management in the United States*: *The Board's Perspective*. Arlington, VA: Nuclear Waste Technical Review Board.

Cohen, B. (1990). *The Nuclear Energy Option*: *An Alternative for the 90's*. New York, NY: Plenum Press.

Compton, A. H. (1956). *Atomic Quest*. Oxford, U.K.: Oxford University Press.

Covello, V. T. and Mumpower, J. (1985). "Risk analysis and risk management: An historical perspective." *Risk Analysis* 5(2):103–20.

Dudley, H. C. (1975). "The ultimate catastrophe." *Bulletin of Atomic Scientists* 21–4.

Fetter, S. and von Hippel, F. (1990). "The hazard from plutonium dispersals by nuclear-warhead accidents. *Science & Global Security* 2:21–41.

Flynn, R., Kasperson, R., Kunreuther, H., and Slovic, P. (1997). "Redirecting the U.S. high level nuclear waste program." *Environment* 29:3, 6–11, 25–30.

Grier, B. (1981). "The early history of the theory and management of risk." Paper presented at the Judgment and Decision Making Group Meeting, Philadelphia, PA.

Kammen, D. M., Shlyakhter, I., and Wilson, R. (1994) "What is the risk of the impossible?" *Journal of the Franklin Institute* 331A:97–116.

Kerr, R. (1996). "A new way to ask the experts: rating radioactive waste risks." *Science* 274(8 November):913–14.

Kleindorfer, P., Kunreuther, H., and Schoemaker, P. (1993). *Decision Sciences: An Integrative Perspective*. New York, NY: Cambridge University Press.

Morgan, M. G. and Henrion, M. (1990). *Uncertainty: A Guide to Dealing with Uncertainty in Quantitative Analysis and Policy Analysis*. New York, NY: Cambridge University Press.

Mossman, K. L., Goldman, M., Mass, F., Mills, W. A., Schiager, K. J., and Vetter, R. L. (1996). "Health Physics Society position statement." http://www.sph.umich.edu/group/eih/UMSCHPS/hprisk.htm, accessed July 7, 1998.

NASA. (1995). *Final Environmental Impact Statement for the Cassini Mission.* NASA, Solar System Exploration Division, Office of Space Science. (June.) Washington, DC. Quoted in Karl Grossman. (1997). *The Wrong Stuff*. Monroe, ME: Common Courage Press.

Nero, A. V. (1979). *A Guidebook to Nuclear Reactors*. Berkeley, CA: University of California Press.

Newsweek. (1994). "Rating the risks by carrier at home and abroad." September 19, p. 22–3.

Oppenheim, L. (1977). *Ancient Mesopotamia*. Chicago: University of Chicago Press.

Shlyakhter, A. I. (1994). "An improved framework for uncertainty analysis: accounting for unsuspected errors." *Risk Analysis* 14:441–7.

Starr, C. (1969). "Social benefit versus technological risk." *Science* 165(19 September):1232–8.

U.S. Nuclear Regulatory Commission. (1975). *Reactor Safety Study: An Assessment of Accident Risks in U.S. Commercial Nuclear Power Plants. Executive Summary, WASH-1400*. NUREG-75/014. Washington, DC.

———. (1986). *Fault Tree Handbook*. NUREG-0492. Washington, DC.

von Hippel, F. N. and Cochran, T. (1991). "Chernobyl: estimating the long term health effects." In *Citizen Scientist*. New York, NY: Touchstone Books.

9

Decision Making

Introduction

The ultimate goal of risk analysis is informed decision making. This requires that information and values be wedded in a thoughtful and consistent fashion, balancing quantitative and qualitative information, diverse priorities and perspectives, and uncertainty. Ideally, the role of models in decisions is to describe the issue, including both what is known and what is unknown. To inform decisions effectively, models must characterize risks in a fashion that incorporates expected end-points, known variability, sources and effects of uncertainty, and implications of assumptions. The preceding chapters developed tools with which to produce information for risk characterizations; this chapter integrates that material and presents formal and informal tools for comparing and selecting among alternatives.

The following problems are based on real and realistic decision situations, some of which you have seen earlier in this book. In some cases, the dominant policy alternatives will be clear, but in others, the choices will not be so clear, for several reasons. One possibility is that uncertainty will be so great that, even if all interested parties agree on valuations, attitudes toward risk taking will lead to contention. Another possibility is that different individuals will disagree about the validity of the assumptions. Yet another is that, even with low uncertainty and little dispute about the science, valuations of outcomes will be dramatically different. Nonetheless, the decision process will be made no worse by risk analysis that is transparent and uses straightforward assumptions, and at best analysis can clearly distinguish among inputs and potential outcomes.

Problem 9-1. Comparing Risk Reduction Measures By Dollar Values

Tengs et al. (1995) reviewed several hundred published analyses of the costs of a variety of life-saving interventions from diverse federal programs. They found that the direct costs per life-year saved ranged from tens of billions of dollars to negative values (i.e., overall gains to the economy). Table 9-1 summarizes the median of cost per life-year saved estimates as a function of sector of society and type of intervention, and table 9-2 contains a cost comparison of a set of interventions.

a. In general terms, what do these numbers mean, and what do or can they tell us? To what extent should we believe them?
b. What does it mean that intervention B costs 99×10^9 per life-year saved? Who is paying this amount?
c. Compare interventions B and C; assume that the two types of facilities are subject to the same regulatory standard. Do you think that the difference in cost between the two was considered before the decision to set the standard? Should it have been?

Table 9-1 Intervention Type by Sector—Median Cost Per Life Year Saved

| | Type of intervention | | | |
Sector of society	Medicine	Fatal injury reduction	Toxin control	All
Health care	$19,000	N/A	N/A	$19,000
Residential	N/A	$36,000	N/A	$36,000
Transportation	N/A	$56,000	N/A	$56,000
Occupational	N/A	$68,000	$1,400,000	$350,000
Environmental	N/A	N/A	$4,2000,000	$4,2000,000
All	$19,000	$48,000	$2,800,000	$42,000

Source: Tengs et al. 1995.

Table 9-2 Selected interventions

	Life-saving intervention	Cost /life year
A	Mandatory seat belt use law	$98
B	Chloroform private well emission standard at 7 papergrade sulfite mills	99×10^9
C	Chloroform private well emission standard at 48 pulp mills	$25,000
D	Cervical cancer screening every 3 years for women age 65 +	≤ $0
E	Smoking cessation advice for men age 50–54	$990
F	Smoking cessation advice for women age 35–39	$2,900
G	Termination of sales of three-wheeled all-terrain vehicles	≤ $0
H	Benzene exposure standard of 1 (vs. 10) ppm in rubber and tire industry	$76,000
I	Radon remediation in homes with levels ≥ 21.6 pCi/l	$6,100

    d. Compare the two negative-cost (i.e., benefit) interventions (D and G). If both of these will have a net positive effect on the economy and save lives, should the government require both? Why have a number of states chosen to take action A, but not G?

Solution 9-1

Solution 9-1a

Tengs et al. set seven criteria for themselves when compiling this study:

1. cost-effectiveness estimates are in "cost per year of life saved"
2. costs and effectiveness should be evaluated from the society perspective
3. costs should be "direct" (vs. "indirect")
4. costs and effectivenesses should be net
5. future costs and life-years saved should be discounted at 5%
6. cost-effectiveness ratios should be marginal or "increment" with respect to a well-defined baseline
7. costs are converted to 1993 dollars

Even given such thoughtful criteria, finding a common metric in which to express diverse risks is a contentious issue for a number of reasons.[1] Deciding what to do is not clear, even given such diverse cost differences as those identified here. Some of these risks are taken by individual, others are involuntary. Some of the risks include other, often nonquantifiable benefits (a significant limitation of the "no indirect cost" criterion). Still others require government intervention that would strike many Americans as intrusive. Nonetheless, the extreme differences suggest that some more efficient risk management options are plausible and may be preferable.

Tengs et al. acknowledge that there are significant uncertainties in the estimates, uncertainties that are not explicit in their summary table. By now, some of the sources of that uncertainty should be familiar; for example, we have seen how different assumptions about the toxicity of benzene can lead to substantially different risk estimates. Clearly, some of the estimates here are conservative, while others are central values. Yet even if these are point estimates representing several order of magnitude ranges, there remain dramatic differences, especially at the extremes.

---

[1] For example, criterion number 2 pits the benefits to an idealized collective against those to a very real, specific individual.

Solution 9-1b

Clearly, no individual company is spending 99×10^9—the entire U.S. pulp and paper industry is not worth this much. Instead, the industry has been required to meet a preset standard of chloroform emissions, and the equipment and process changes needed to meet that standard cost some (significantly smaller) amount. The life-years saved are an expected number based on estimated human exposure to chloroform and some dose response associated with that exposure. The cost of controls divided by the fraction of an expected life-year saved gives the 99×10^9 value.

Solution 9-1c

It is unlikely that the difference in cost among individual facilities was considered. Rather, a single emission standard was set for all of the industries—this is referred to as a "command and control" regulation. This type of regulation has the advantage that it is clearly enforceable and achieves the immediate objective, but it is often decried for failing to make efficient use of resources.

In contrast, a regulatory approach that bases controls on economic efficiency can ignore some important social implications. Setting a cost per life-year saved target and allowing different emission levels can mean that those who are expected to bear the decreased lifetimes face different risk levels. For example those close to a highly polluting facility will be subject to higher risk relative to those near a cleaner facility, while neither group gets any of the economic rewards of the operation. This discrepancy between those who pay abatement costs and those who bear the risks is one reason that many environmental and workplace regulations are based on set risk levels, rather than cost per life saved estimates.

Solution 9-1d

Several quite difference inferences can be drawn from this type of comparison, with quite different policy implications. It is possible that interventions with net economic and life-saving advantages have not been taken because the benefits were not known prior to the analysis, because they do not fit clearly into an existing regulatory category, or because of institutional inertia. On the other hand, it is possible that a broader set of values informs these issues, including nonmonetarizable effects and the indirect costs explicitly omitted from this study.

The greatest value of comparing along a common metric is that it provides a framework from which to decide whether policy choices are poorly prioritized or whether the analysis provides an incomplete picture of the situation. The Tengs et al. study suggests both that a number of existing regulatory policies merit careful scrutiny and that using "dollar per life-year saved" is an insufficient absolute decision criterion.

The cervical screening case indicates that preventative measures can present immediate costs while providing significant long-run savings. Medical insurance groups and health maintenance organizations recognize this and can profit from it; for noninsured individuals, or cases where the benefit is too long term to meet private financial goals, government intervention can provide or encourage such actions.

The all-terrain vehicle (ATV) case suggests that riding these vehicles is a significant risk—and probably one borne entirely by the ATV riders themselves. Should the government be involved in this type of decision? This is clearly a question that analysis can inform, but it cannot provide an "answer." Rather, politics must provide answers that are informed by a broader set of values than just direct economic costs and a one-dimensional measure of benefits.

Problem 9-A. Cost Per Life-Year versus One in a Million

Compare this measure of risk to those used by Wilson in table 1-1. What are the similarities? Differences? How might each be used to set policy?

Problem 9-B. Too Much or Too Little Spending?

Consider table 9-1. Based on this, are we spending too much on environmental toxin control? Too little on health care?

Problem 9-C. Mandatory Helmet Laws

Mandatory helmet laws for motorcycles also represent a negative-cost intervention. Compare this to interventions D and G.

Problem 9-D. Discussion

Discuss possible reasons for and the significance of the difference between interventions E and F.

Problem 9-E. Is It Worth It?

The life-years saved by intervention H would be among employees, and consequently it is an OSHA decision, which means that cost can be considered. Is it worth it? Why or why not?

Problem 9-F. High-Level Nuclear Waste

Suppose Tengs et al. had found and included an analysis of the cost per life-year saved for various high-level nuclear waste disposal options, including

a. separating the plutonium and using it for power in breeder reactors
b. temporarily storing it on site, and
c. storing it permanently underground.

Evaluate and discuss whether, how, and why each of the combinations of costs in table 9-3 would affect your preference for each of these options.

Organizing Processes

A variety of approaches have been developed to organize and depict processes. Problem 9-2 revisits "fault trees," and demonstrates how they

Table 9-3 Hypothetical Cost per Life-Year Saved Scenarios for Problem 9-F

	a	b	c
i	≤ $0	$20 × 10^6$	$4000
ii	$10,000	$10,000	$10,000
iii	$20 × 10^6$	$1000	$20 × 10^6$
iv	$0–20 × 10^6$	$1000–10,000	$100,000– 1,000,000

can help guide decisions about a process. It also includes some caveats about their application to highly complex situations.

Problem 9-2. Event Trees and Decision Analysis

Consider the fault tree in figure 9-1 for a process and a manufacturing facility. Assume that a successful operation is worth $500 to the company, an unsuccessful operation costs the facility $500, and an unsafe

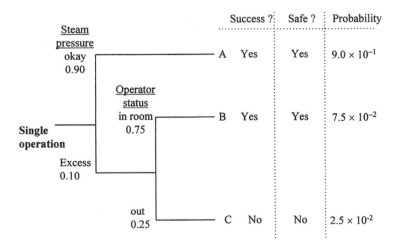

Figure 9-1. Event tree for problem 9-2a.

operation costs an additional $20,000 (damage to process equipment and repair of the valve). In addition, it costs the facility $350 for each operation in general expenses, regardless of success and safety, and it needs to average $50 in profits per operation to remain competitive.

a. Calculate the expected value to the facility of an operation.
b. How much should the facility be willing to pay to install a 90% effective pressure relief valve?
c. Suppose it would cost $80 per operation to hire an additional operator, thereby insuring ($P = 0.99$) that at least one operator would see the warning light. Assume further that this would not make the operation more profitable, and that a safe but unsuccessful operation would still cost the facility $500. How much would an unsafe operation have to cost in order for the expected value to the facility to just meet its target?
d. Assume that the facility has one operator and has installed a 90% effective pressure relief valve as in (a), and that the original outcome values apply. Suppose that an unsafe operation invariably forces the evacuation of a warehouse next door as a cloud of hydocarbons and catalyst is released. This costs the warehouse owners $14,000 in lost work hours and other costs. What is the value of the operation to the facility? To the warehouse owner? What is the overall expected value of the operation? How might the company and the warehouse owners deal with this situation?
e. Suppose that instead of the conditions in (d), the cloud blows toward the warehouse only 10% of the time. In addition, 60% of the time the wind blows the cloud over an unpopulated area, and the final 30% of the time it blows across an area populated by about 1000 people. A combination of exposure and dose-response data suggests that 3×10^{-4} life-years will be lost in the population in general with each such event and on average one in every hundred thousand persons will have an acute, fatal allergic reaction to the catalyst, which is not detectable a priori.

i. What is the overall value of the operation now?
ii. How could the company, the warehouse owners, and the local residents deal with the situation?
iii. Could or should there be a role for government in this situation?

Solution 9-2

Solution 9-2a

The expected value of a single operation is the sum of the values of each individual outcome times the probability that the outcome will occur, minus other costs (expenses and profits). This can be written as

$$\text{expected value} = \sum_i [P_i \cdot V_i] - Q, \qquad \text{9-1}$$

where i is the possible outcome, P_i is the probability of that outcome, V_i is the value of that outcome, should it occur, and Q is all other costs. In this case, without a safety valve, there are two possible outcomes: successful and safe (A and B), and unsuccessful and unsafe (C), so

$$\text{expected value} = (0.90 + 0.075) \times \$500$$
$$+ (0.025) \times (-\$20,500) - \$400$$
$$= -\$425.$$

In other words, the facility can expect to lose on average of $425 per operation, so the company should choose not to run the operation.

Solution 9-2b

How much should the facility be willing to pay to install a 90% effective pressure relief valve?

See figure 9-2 for the event tree in this case. Plugging the values into equation 9-1, we find that with the pressure relief valve (PRV),

$$\text{expected value} = (0.90 + 0.075) \times \$500 + (0.023)$$
$$\times - \$500 + 0.0025 \times (-\$20,500) - \$400$$
$$= \$25.$$

The facility should be willing to pay $25 per operation for the pressure relief valve.

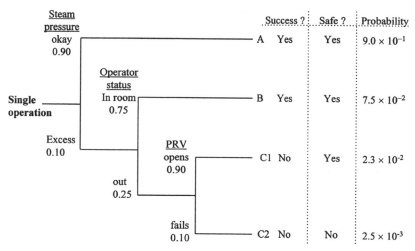

Figure 9-2. Event tree for problem 9-2b.

Solution 9-2c

In this case, we know the probabilities, expected values, and most of the outcome values and are solving for one of the individual values. Now, however, we need to add $80 per operation to Q:

required expected value = $0;

$$\$0 = (0.90 + 0.099)$$

$$\times \$500 + 0.0001 \times (V_c) + 480;$$

$$V_c = \frac{\$480 - (0.999 \times \$500)}{0.0001}$$

$$= -\$195,000.$$

If the cost of an unsafe operation is less than $195,000, it does not make sense for the company to hire the additional operator.

Solution 9-2d

We already calculated the expected value to the facility in (b): $25.
    The value to the warehouse is $0 for any safe operation, and −$14,000

for an unsafe operation. Since the probability of an unsafe operation is 0.0025 (see figure 9-2),

$$\text{expected value (warehouse owner)} = -\$14,000 \times 0.0025 = -\$35.$$

The warehouse owner expects to lose $35 per time the facility next door runs an operation; in economic terms, this is a $35 negative externality.

To calculate the overall expected value, we could go back to equation 9-1 and substitute $34,500 for the overall cost of an unsafe operation. However, the result will be the same as the sum of the two individual expected values, so

$$\text{expected value (overall)}$$
$$= \Sigma(\text{individual EVs}) = \$25 - \$35 = -\$10.$$

Thus, there is an overall negative expected value associated with each operation. How the two businesses might deal with this is not obvious without some additional assumptions:

- If property rights are clearly defined and include this type of emission, the operations would cease.
- If the facility had a generally accepted right to have some unsafe operations, the warehouse could pay it $25.01 for each time it did *not* run the operation. Under this scheme, the facility would make 1 cent more, and the warehouse would lose $9.99 less, for each (non)operation.
- If, on the other hand, the warehouse owner had a right to be compensated for each evacuation, the facility would not be able to afford such compensation, and would choose instead not to operate.

In many cases, property rights are not so clearly defined, and trying to resolve the issue in the courts can entail significant additional costs (what economists refer to as "transaction" costs). Alternatively, the government might regulate one or both facilities, by fining the facility for unsafe operations, for instance, or requiring additional safety equipment.

Solution 9-2e(i)

Now we begin to see how complicated this sort of decision becomes, even when simplified to the level of three "actors" and one operation (and without uncertainty!). For many urban areas, there are hundreds of thousands of "actors" and operations. Computing an "overall value" becomes difficult, if not impossible. The values to the facility ($25) and the warehouse [(0.1 × − $35) = − $3.50] are straightforward. But can we tell whether the value to the community is more or less than $21.50, the break-even amount? If so, how?

One approach, suggested in problem 9-1, is to calculate what the value per life-year lost would need to be for the overall value to break even. There is a 2.5×10^{-3} chance of an unsafe operation, and a 0.3 chance of community impact when there is such an event. For toxic effects, the outcome is 3.0×10^{-4} life-years lost. Assuming that a fatal allergic reaction will take 35 life-years with an expected probability of 1×10^{-5}, the expected outcome is another 3.5×10^{-4} life-years lost, for a total of 6.5×10^{-4} life-years lost per event. Consequently, the social break-even value for lost life-years is

$$V(\text{life-year}) = \frac{\$21.50}{0.3 \cdot (6.5 \times 10^{-4} \text{ life-years}) \cdot (2.5 \times 10^{-3})}$$

$$= \$44 \times 10^6 \text{ per life-year lost.}$$

Solution 9-2e(ii)

If all parties could agree on property rights and on a dollar value per life-year saved, and if they agreed on all of the analysis that led to the expected-value calculations, an "economically efficient" solution could be achieved. For example, what if an agreed-upon value per life-year lost were less than 44×10^6, and the rights were determined to lie with the warehouse and the community? In this case, the facility could pay warehouse evacuation costs, pay some amount ≤ $21.50 to the community for each operation, and still make an acceptable profit.

Experience suggests that this is not a likely outcome.

Solution 9-2e(iii)

To force agreement, in many cases government intervention is the only viable solution. Indeed, if there were no government in the area, the community might well establish one to deal with such a situation!

- If a government could determine the values and rights, it could impose taxes or penalties in an economically efficient fashion, with the government acting as an agent that reduces transaction costs. This solution is appealing at some level, but in practice is frequently impossible to implement and has significant philosophical shortcomings.

- If, on the other hand, the rights of the citizens to safe air were considered inviolable, a government could require the facility to stop operations. This is a major issue in worldwide debates about regulation and especially the role of cost-benefit analysis. Many economists and corporations believe that if the net social benefit of production, including environmental externalities, is positive, then the overall economy benefits and the operation should be permitted. Others argue that a cost-benefit approach allows some individuals within the economy to profit at the expense of others, and that lower overall economic gains in a more equal world can be a superior outcome.

- If the determination of rights and economic values were not clear, the government might establish a maximum level of risk to which the facility could subject its neighbors. This solution dominates environmental regulation, with recurring 10^{-X} risk standards (especially where $X = 5$ or 6).

- Similarly, a government could require a given set of controls, such as backup PRV's or redundant operators. This type of "command and control" solution is often based on achieving on average a 10^{-X} outcome. The command and control approach is the most arbitrary in term of achieving a specified outcome, especially if there is significant diversity among different facilities. For example, one company might be able to install a backup PRV system for $10 per operation, while it would cost another $400; a command and control approach would not distinguish between the two. It is this type of differential that creates discrepancies such as those seen in table 9-2, items B and C. Nonetheless, this approach is often the easiest to establish and enforce, and does achieve results.

Clearly then, the role of government is a hybrid of interests, philosophies, and economic needs. The analysis can inform decisions, establish some parameters, and identify possible solutions, but it cannot determine underlying rights, beliefs, and values; instead, these must determine the analysis.

Problem 9-G. Who Lives There (and Does It Matter)?

Problem 9-2d did not give any information about the 1000 people living in the affected area.

a. How might your appraisal of the situation and the alternative differ if the area were populated by

    i. a balanced cross section of the U.S. population?

    ii. 1000 middle-class individuals, most with high-school diplomas and about half with bachelor's degrees, all of whom use the type of product made in the process?

    iii. individuals as in (b), except that only fifteen percent use the type of product?

    iv. individuals who work at the facility and their families?

    v. low-income minorities who regularly use the product?

    vi. low-income minorities who never use the product?

b. Assume that the residents are a balanced cross section of the U.S. population. Compare how their perceptions of the situation might vary if

    i. both they and the facility have been in the area for many years, and the secondary valves were installed five years ago,

    ii. the facility has been around for many years, and installed the secondary valves five years ago, but most of the residents moved in over the past five years,

    iii. the residents have lived in the area for many years, but the facility was installed five years ago.

Problem 9-H. Additional Calculations

Suppose that, given an unsafe operation, there is a 30% chance that the cloud will drift through the community, lingering for one to three hours.

a. If the facility is operating as in figure 9-2, with the addition of secondary safety valves with 2.5×10^{-5} chance of failure per use, has 10 such units and runs each unit every 2 hours, eight hours per day, five days per week, how many times will a cloud pass through the community each year?

b. Suppose that detailed analyses indicate that the releases are non-toxic. Nonetheless, residents complain that the resulting odors cause headaches, runny eyes, and mild nausea, as well as disrupting their lives (for example, decreasing the enjoyment of meals and causing stress among residents who don't believe the odors are nontoxic). Could these complaints be incorporated into an equation 9-1 type of valuation? If so, how? How else might the community (including or excluding the facility) deal with the situation?

Problem 9-I. Additional Uncertainty

Problem 8-3 discussed some major sources of uncertainty for this type of analysis. Posit a plausible scenario for each of the three categories (missing component, common-mode failure, and human element) for figure 9-2. Evaluate numerically and discuss the implications.

Problem 9-3. Analysis or Abuse? Transmission of Hoof Blister

Consider the following *hypothetical* situation. An animal health agency (AHA) asked a consultant to evaluate the risk of transmission of porcine hoof blister (PHB) into the domestic pig stock from foreign "quarantine swine," or pigs that are brought into the United States to be fattened and butchered. PHB is a disease that causes symptoms in pigs such as fertility problems and markedly reduced body weight, and can be transferred to humans, where it can cause flulike symptoms. The disease has never been found in domestic pigs, so risk of transmission from abroad is of concern.

Several measures are in place to insure that the quarantine swine are not infected. A vaccination for PHB exists, and all imported pigs must either show certification of vaccinations or be given blood tests at the port of entry, where infected pigs are destroyed. In addition, all pigs' hooves and skins are inspected at the border for characteristic visible symptoms of PHB and other diseases. Nonetheless, it is possible that infected pigs will escape detection if they have what are called "latent" infections, which can occur when piglets are infected in utero, through mother's milk, or from some other oral source prior to vaccination, but the infection is not active.

The consultant had the following (very limited) data to work with:

- The frequency of PHB in foreign pig farms has been estimated by two different researchers as 0.5% and 7.2%. No other studies are available.
- The chance that a piglet from an infected sow will acquire a latent infection is between 2.5 and 10%, based on several studies.
- The number of quarantine swine that entered the United States in 1998 was known to be 51,224.

This problem is loosely based on a real analysis that was submitted to a real agency. It suggests how overanalysis and inappropriate use of tools can lead to spurious certainty and create an artificial boundary between experts and nonexperts. It also shows how an understanding of analytical tools can allow one to sort through a completed analysis and get at the underlying assumptions and embedded preferences.

   a. What is the probability that a quarantine pig entering the country has a latent PHB infection?

   b. What is the expected number of infected pigs brought into the country each year? The range?

   c. Use Monte Carlo analysis to estimate a range of the number of pigs with latent infections that might have entered the country in 1996. Explain the reasons for choosing the distributions. Repeat the analysis with other possible distributions and discuss the differences. Does Monte Carlo analysis provide better information than was calculated in part b?

Quarantine swine are kept separate from pigs in domestic farms. However, the consultant found a number of possible points of contact that could lead to infection, such as piglets born to quarantine animals after importation and then placed with domestic pigs, contact between mature animals in an "animal hospital," where sick pigs from both stocks may be pooled, and contact by a domestic animal with bodily fluids from an infected animal in various common areas. With no available information on numerical probabilities for the risk of transfer to domestic animals, the consultant used a qualitative technique, in which each parameter is given a numerical estimate, and is then multiplied by a factor indicating the certainty of that estimate. Specifically, he made the following assumptions (using only qualitative support information):

- the likelihood of transmission gets a "medium rating," numerically valued at 0.5
- the degree of certainty associated with this rating is "moderately" certain, which is numerically valued at 0.75
- the consequence of transmission if it occurs is given a "high" rating, which is numerically valued at 0.01
- the certainty of the consequence estimate is "certain," which is numerically valued at 0.5

From this he calculated the expected risk associated with transmission as

$$E(\text{risk}) = [(\text{likelihood}) \times (\text{degree of certainty})]$$
$$\times [(\text{consequence}) \times (\text{degree of certainty})]$$
$$= (0.5 \times 0.75) \times (0.01 \times 0.5)$$
$$= 0.001875.$$

d. Evaluate this calculation. What does the number mean? Does it provide better information than the narrative that led to the estimate?

The consultant concludes that "[t]he risk of exposure and spread within and out of the ... feedlot is high," and also argues that "foreign pig interests will require the AHA to back any decisions regarding this issue with sound science and fact."

   e. If you were a foreign pig importer, or a government official from one of the exporting countries, how would you feel about the conclusion that there is "high" risk of transmission? If you were a domestic pig farmer who did not raise quarantine swine, and were concerned about your pigs contracting PHB in this way? Does there appear to be sufficiently "sound science and fact" upon which to make a decision?

Event-decision trees are a tool for organizing complex decision scenarios such as this one. They allow the analyst to subdivide the problem into discrete pieces, and identify the range of possible outcomes. Once the outcomes are identified, they can be assigned values. In many cases, the valuation need not even be numerical; rather, some outcome or subset of possible outcomes may be identified as clearly dominant.

   f. Generate an event-decision tree for this problem. Discuss how it might help clarify this problem. On which key points are the two sides in this debate likely to agree and disagree? Does the analysis in parts a–d inform the "important" issues?

Solution 9-3

Solution 9-3a

This probability can be calculated relatively easily given the values, but clearly there is a fair amount of uncertainty. The probability that a pig has the disease is calculated as the probability that its mother is infected times the probability that the disease is passed on if the mother is infected, or

$$P(\text{diseased offspring})$$
$$= P(\text{diseased mother}) \times P(\text{transfer to offspring}).$$

Since there are high and low estimates of both input values, this probability is best characterized by a range {low, high}:

$$\text{Low: } P(\text{diseased offspring}) = 0.005 \times 0.025 = 0.0001;$$

$$\text{high: } P(\text{diseased offspring}) = 0.10 \times 0.072 = 0.007 \approx 0.01.$$

Thus the already highly uncertain information yields a two orders of magnitude range of the possible fraction of diseased swine, {0.0001, 0.01} or {0.01%, 1%}.

Solution 9-3b

It is not clear what the expected value is, since there is no clear estimate of the probability of infection. Without better information, the expected value of the probability might be taken as 0.001, but the range is probably a better representation. The range of possibly infected swine is $P(\text{infected}) \times$ number in herd, or from

$$0.0001 \times 50,000 \text{ swine} = 5 \text{ swine (low)}$$

to

$$0.01 \times 50,000 \text{ swine} = 500 \text{ swine (high)}.$$

Solution 9-3c

The Monte Carlo inputs are the probability that a sow is infected and the probability of transmission. Since very little is known, uniform distributions might be considered appropriate in both cases. Ten-thousand iterations using

$$P(\text{infection}) - \text{uniform } (0.005, 0.072),$$

$$P(\text{transmission}) - \text{uniform } (0.025, 0.10),$$

$$\text{number in herd} = 50,000$$

yields a 95% confidence interval of {17.50, 302.17}, which by now you should recognize as identical to {20, 300}.

Contrasting this with the range based on extremes, {5, 500}, suggests that Monte Carlo analysis provides no additional information. Other input distributions provide different outputs, but are of the same order and are thus not really "better" outputs.

Solution 9-3d

This is a very problematic use of probabilistic risk assessment, and represents a case of trying to use a tool that is inappropriate for the question. Assigning numerical values based entirely on subjective interpretations of aspects of a risk might have some use for rough comparisons, but a precise (four significant figures!) result with neither context nor units implies a level of understanding that is not warranted by the available information. In addition, the whole original calculation of the number of infected animals is irrelevant to the "real" question of transmission to domestic stock, except insofar as it shows that transmission is possible. The argument that transmission to domestic stock is possible stands, but is in no way improved by a subjective risk estimate.

Solution 9-3e

It seems unlikely that the foreign importers would be satisfied with this sort of analysis. In fact, the process that the consultant followed essentially argued that "the risk is high, and therefore it is a high risk." Someone on the other "side" of this issue could easily conclude that the assessment was at best poorly structured and at worst intentionally designed to appear frightening.

This problem illustrates how easy it is to make an analysis appear scientific, when in fact it is an entirely subjective process. Another assessor might make the alternative assumption that the likelihood of transmission deserves an "impossible" rating, numerically valued at 0.00, in which case the other parameters are unimportant, since the product will always be 0.

On the other hand, there does appear to be some legitimate concern on the part of domestic pig farmers, a concern that cannot be dismissed by arguing that the assessment is inadequate. The process used in (f) helps illuminate this issue.

Solution 9-3f

Decision analysis is a formal, structured approach to decision making, often of great value for highly uncertain decisions. It allows a decision maker to evaluate a problem in broad terms first, which can often lead to dominant ("best") choices even without detailed analysis. In addition, if dominant decisions are not clear, this framework can indicate where additional analysis will be most useful. Depictions such as figure 9-1 should be generated at the beginning of any decision process.

The following is a useful approach for structuring decision problems (after Kleindorfer et al., 1993):

1. Identify the nature of the problem, including all alternatives and objectives
2. Evaluate the impacts of the alternatives, specifying uncertainties and consequences
3. Determine preferences of decision makers, including value trade-offs and diverse objectives
4. Evaluate and compare alternatives by developing alternatives and performing sensitivity analysis
5. Select the best choice

A first cut at this problem might look at only one objective, avoiding transmission of PHB to the domestic stock. In fact, this appears to be what the consultant did consider, albeit in a less formal fashion.

Alternatives. A variety of alternatives exist, but these can be simplified to three, two extremes and one intermediate. One extreme (call it $A1$) is to maintain the status quo. For $A1$, there is some probability (call it $P1$) of transmission. The intermediate ($A2$) is to impose some additional restrictions. For $A2$, there is also some probability of transmission, $P2$, where $P1 > P2$. The third alternative ($A3$) is to ban exports. For $A3$, the probability of transmission is 0.

Impacts. In this scenario, there are two impacts. One is that transmission occurs, in which case there is some cost (C) to dealing with the disease. The other is no transmission, in which case costs are 0. Expected values will vary, however, as the probabilities vary.

This leads to the decision structure shown in figure 9-3. Squares refer to decision nodes. Circles refer to event nodes, which lead to possible outcomes given a decision. Outcomes are beyond the control of the decision maker. On the basis of this tree, both of the no-ban options result in some probability of incurring costs, while the ban option results in no costs. If this analysis captures the whole picture, the ban is clearly the dominant decision, and it is not even necessary to calculate $P1$ and $P2$. Note, however, that there may be other sources of transmission even given the ban, such as illegal trade.

Another approach is to look at a wider range of objectives. The first has already been proposed: to avoid transmission of PHB to the domestic stock. Another is the additional indirect cost of reducing or eliminating imports, which means that domestic consumers will be

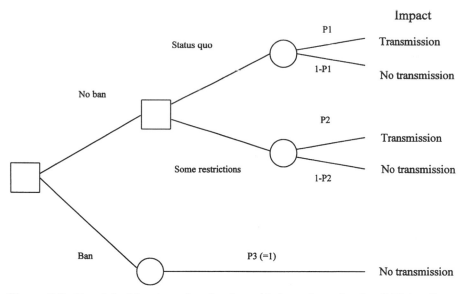

Figure 9-3. Event-decision tree for the transmission of porcine hoof blister from quarantine swine to the domestic pig stock, based on data from problem 9-3.

paying higher prices for pork. A third possible cost might be the more difficult to evaluate, but nonetheless important, desire of domestic farmers to feel protected from a highly uncertain situation. Finally, there is the desire of foreign interests to continue exporting pigs at the lowest possible cost, and possible additional trade issues, such as treaties and impacts on other markets. Using decision analysis, the consideration of values becomes explicit.

Problem 9-J. Dollars and Decisions

a. Assume that the only alternatives are a ban and the status quo, that the probability of transmission with no ban is $P1 = 0.01$, that the cost associated with transmission is $C = \$1,000,000$, and that the only other costs are the indirect costs paid by domestic consumers. Draw a decision tree for this scenario. How much would the indirect costs (I) have to be to make the decision maker indifferent between the ban and no-ban options?

b. Repeat (a), using $C = \$1,000,000$ and $P1 = \{0.002, 0.05\}$. Would there be a clear decision preference if $I = \$1000$? If $I = \$5000$? If $I = 50,000$?

c. Create a decision tree using alternatives A1–A3 and the four costs discussed in the paragraph above. Discuss whether and how this tree would be useful to a decision maker from the AHA. Do the costs need to be measured and compared? If so, suggest how to go about it.

d. Sensitivity analysis. Review the problem and suggest and defend some alternative values of the Ps. Which points in the decision process, if incorrect, are most likely to affect final decisions?

Problem 9-4. Health and Environmental Technology Policy: Superfund Remediation

Situation Description. An abandoned chemical waste dump, adjacent to a low-income residential community, is contaminated with polychlorinated biphenyls (PCBs) and the heavy metal cadmium, both of which are known to cause, at very low exposure levels, a wide variety of toxic

effects, both carcinogenic and noncarcinogenic, in humans. Other contaminants may be present but have not been identified. The site has
been targeted for remediation under CERCLA (the Superfund Act).
Two possible approaches to the cleanup have been identified by the
EPA and are under consideration:

A1. Excavation of waste matter, on-site incineration, and reburial of
    ash in a lined, on-site, landfill in corrosion-resistant drums
A2. In situ (in place) stabilization of waste by erecting hydrologic
    barriers around the site, together with the pumping out and
    treatment of contaminated ground water

U.S. Environmental Protection Agency Risk Assessment. EPA risk assessors have evaluated the increase in individual risks of the two
approaches over time, assuming that the best available technology is
employed in both cases. Risk here is defined to be the annual excess
probability of contracting a fatal cancer experienced by the most affected individual as a result of the cleanup operations. The assessment
results are shown in figure 9-4. For comparison, the "do-nothing" curve
is also included. In figure 9-5, the aggregate risk over a one-hundred-year
period is shown for both cleanup methods, as a function of cost.

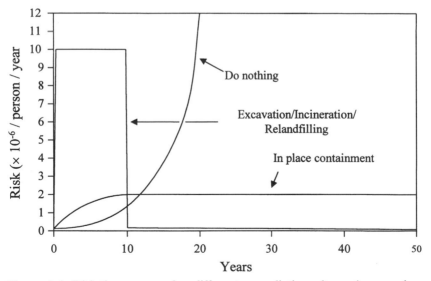

Figure 9-4. Risk-time curves for different remediation alternatives as described in problem 9-4.

a. Based on figures 9-4 and 9-5, determine the major pros and cons for each of three alternatives:

incineration, in-place containment, and doing nothing

b. Identify the most important uncertainties and missing data

Some information about figures 9-4 and 9-5 is supplied:

i. incinerator performance data were supplied by the company that would receive the contract to do the work; no independent verification was performed.

ii. The incinerator is said to destroy 99.99% of the PCBs charged to the incinerator. Cadmium is not destroyed but is partitioned between the ash, the scrubber, and the stack discharge.

iii. Incineration reduces the waste volume by 85% and produces contaminated wastewater from the stack scrubber.

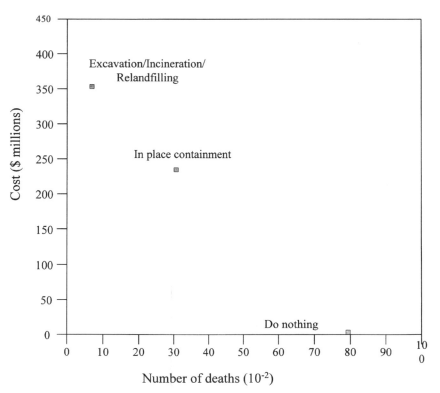

Figure 9-5. Risk versus cost of abatement as described in problem 9-4.

iv. The risk values for the incineration option are estimated to be accurate within a factor of 10; for the containment option, within a factor of 2.

Solution 9-4

This problem is representative of many decisions about environmental issues: decisions must be made for a variety of reasons, but data are incomplete and of questionable quality. That the information exists to inform the decisions at all is a positive sign, but it must be applied carefully!

For example, the information on the products of incineration has been provided by the company that sells the product. How much faith can be put in these numbers? It is unlikely that there will be consensus on the trustworthiness, which can lead to substantially different conclusions. Some of these concerns might be answered by further investigation: are the 85% and 99.99% values ideal, or under real conditions at other locations? What is the track record of this type of equipment? On the other hand, the answers even to further questions may be accepted at face value by some, and assumed to be wrong by others.

In addition, some of the uncertainty terms are quite vague. What do the "factors" of 10 and 2 mean in terms of the two figures? Do they describe a 99% confidence interval? An absolute maximum and minimum? How reliable are they? Figures 9-6 and 9-7 repeat figures 9-4 and 9-5 with one interpretation of the factors from (iv). Do these figures make the decision any easier? Elaborate calculations could estimate the net present value of each option. Given the extent of uncertainty, would this inform or detract from the solution?

The following are some arguments for and against each of the options. Clearly, there is no "exact" solution, but a careful review of the evidence goes a long way toward cutting through the natural arguments of the likely stakeholders (the community, EPA, site owners, and so forth).

Incineration

Arguments for:

- Incineration would be a "permanent" solution since only ash residue would remain from the waste.

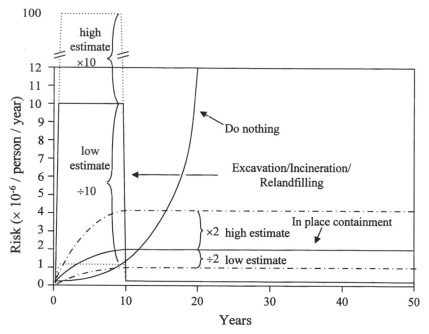

Figure 9-6. Risk-time curves for different remediation alternatives, with uncertainty bars as described in problem 9-4.

- The problem is solved "once and for all." The long-term risks (after incineration is complete) would be very low.
- The physical volume of the refuse would be reduced.
- It is possible to reuse the incinerator in the future.

Arguments against:

- It is the most expensive of the available options.
- The risk is much higher during the decade of operation.
- The uncertainty to a factor of 10 of the risk data for this option (especially since it is provided by the industry, which would certainly not err on the conservative side) may be unacceptable.
- The 99.99% operating efficiency of the incinerator is a dubious assertion at best. Startups, upset conditions, old equipment, poorly trained and inexperienced operators, absentee company officials, insufficient regulatory oversight, weather, and other realities are certain to reduce this operating efficiency.
- The worst-case scenario for incineration (a risk ten times higher than that given) will be in the order of 10^{-4}.

Figure 9-7. Risk versus cost of abatement, with uncertainty bars as described in problem 9-4.

- There will be aesthetic damage to the landscape.
- There is high immediate cost.
- There will be a high concentration of contaminants in the ash.
- The cost per expected statistical life saved (beyond that of in-place containment) is very high (650×10^6). (Note that this could be used as justification for the do-nothing alternative.) However, many programs with price-tags at or above this cost have been implemented by state and federal agencies.

In-Place Containment

Arguments for:

- With a discount rate of 5% for risk, the total risk of in-place containment is always less than that of incineration. With a 1% discount rate the relationship holds for the first 64 years.

- It delays the decision, which is desirable if new disposal schemes are likely in the future.

Arguments against:

- If the site is near a residential area the soil and water contamination may have immediate effects.
- The problem is "still there."
- With no discounting, containment will have a larger long-run risk than incineration.

Doing Nothing:

Arguments against:

- Fitting an exponential curve to the risk of the do-nothing option and obtaining the present discounted value of the risk with a 5% discount rate shows that the do-nothing approach becomes costlier in terms of risk than the incineration option after twenty-seven years. With lower discounting the time lapse will be even less.
- It could be politically suicidal.

"Missing" Data

The information given in this problem provides some general framework for the decision at hand, but it is unlikely to fulfill the needs of the decision makers. While one will never have "complete" information on a question such as this, some additional issues that are certain to be important for the decision-makers follow.

- Distribution of risks. Are some individuals subject to higher risks than others? Does this change from one abatement option to another?
- Other end points. Risks are only given in one dimension, loss of life. Are there also going to be nonfatal end points? What are they? Are the risks greater for children? The elderly? Can the loss of life be narrowed to life-years lost?
- Population of the region. The risk will be perceived as very different if it is spread across a population of tens versus tens of thousands.
- Exact location of the site and incinerator with respect to residential areas, as well as topography, geology, and chemistry of the area. Without knowing how the exposure was modeled, it is hard to know

what may have been excluded and what uncertainties are not represented in the findings.

- Technical specifications of the incinerator. The manufacturer has provided some information, but an independent evaluation of how such incinerators have worked in "real world" conditions would be useful.
- Political climate. Each stakeholder in this incinerator will probably make an assessment of what they would like and what they can expect to get, as well as how this issue fits into a broader set of needs and activities.
- Discounting. Are the costs discounted to current time or over time?

One important but controversial issue in cost-benefit analysis is the application of discount rates for long term and uncertain risks that carry immediate costs. Examples include risks that span a number of years (such as those of hazardous waste remediation, as in problem 9-4), those spanning centuries (such as future climate change resulting from current carbon emissions), and even those spanning millennia (such as long-term storage of high-level nuclear waste).

One argument often proffered is that discount rates similar to those for financial transactions should be used to determine the appropriate investment in risk mitigation measures. This approach implies that the value today of costs that will be borne in the future is lower than the dollar value of those costs. The advantage is that these discount, or interest, rates are well established.

If one needs to spend one hundred dollars now or one hundred dollars in one year, the choice to do so in one year makes sense because the money could be earning interest between now and then. At an interest rate of five percent, the payment in a year solves the problem and results in income of five dollars. Of course, if the choice were between spending one hundred dollars now and one hundred and ten dollars in a year, paying now would be preferred, using a five percent interest rate as the decision criterion.

Discount rates and interest rates are analogous. If one expects to incur a cost in the future, that cost can be transformed into a "present discounted value," which is the value discounted to today of bearing that cost in the future. The mathematics of discounting are straightforward.

Assume the annual discount rate (d) is constant over time. If the cost in the future to address a risk is C (in today's dollars), then the value *today* of that future cost is

$$V = C/(1 + d).$$

Alternatively,

$$C = V \times (1 + d).$$

Over two years, the discount rate is simply

$$V = C/(1 + d)/(1 + d)$$
$$= C/(1 + d)^2,$$

and t years from now is

$$V = C/(1 + d)^t. \qquad\qquad 9\text{-}2$$

If instead the risk will require an annual expenditure, then the annual cost (starting today, $t = 0$) will be

$$V = C(0) + C(1)/(1 + d) + C(2)/(1 + d)^2 + \cdots + C(n)/(1 + d)^n.$$

This can also be written as a continuous equation for a cost at time t as

$$V = C(T)e^{-dT}. \qquad\qquad 9\text{-}3$$

If future costs can be described as a function of time, $C(t)$, then the value today of costs incurred between now and some time T can be found by integrating:

$$V = \int_0^T C(t)e^{-rt}\, dt. \qquad\qquad 9\text{-}4$$

Although the mathematics are straightforward, the decision to use discounting as a decision criterion is not. Morgan and Henrion (1990) warn that the choice of discount rate, even when based on empirical information from other or past cases, is a value judgment, based on a philosophy that the choice in question can reasonably be compared to other choices. Decisions on future expenditures are often evaluated at discount (interest) rates between 3 and 12%, and many analysts apply these same rates to environmental issues and risk management decisions. This can mean assigning very small current costs if a problem will not require action for a long time. For example, discounted at 6%, a projected expense of one million dollars in ten years is valued at just over half that amount ($558,659) if payment were required today.

Dasgupta and Mäler (1994) argue that for environmental resource decisions in developing countries, some reasonable conditions can suggest that negative discount rates be used. Issues of intergenerational equity and fairness, social and individual preferences, and technological change, for example, are difficult if not impossible to capture in a single number.

Problem 9-K. Discounting or Dodging?

Assume that a policy decision determines that a human life is worth one million dollars. This implies that if a company can expect to save one life by spending one million dollars or less, it should do so. Suppose a company has found it can expect to avoid two deaths if it spends two hundred and fifty thousand dollars on a mitigation measure. However, it points out that the individuals affected are working age now. The deaths avoided are not happening today, but instead are premature deaths in thirty years. This is because two workers are expected to die in their seventies with the mitigation, and in their sixties without it. Consequently, the company asks that the agency carrying out this policy discount the values of these lives over the intervening thirty years.

a. What discount rate would make the cost per life saved in thirty years as discounted to today just equal to the two hundred and fifty thousand dollars that the company would have to spend today?

b. At a discount rate of five percent, what is the value today of a life lost in thirty years?

c. Assume that you want to estimate how much the United States might want to spend on carbon emission reductions today in order to avoid costs in the future. For simplicity, assume that one set of options would save the world economy one billion dollars a year for fifty years, starting in the year 2100. What is the net present value of those savings at a discount rate of ten percent? Five percent? One percent?

d. Keeping in mind that many national policies regarding risk deal with potential future losses, discuss the policy ramifications of your answers to parts a–c. What are some arguments for and against discounting future loss of lives or economic costs? When you read the first line of this problem, did the idea of discounting the value of the lives occur to you? Do you think it occurred to the lawmakers who established the policy? Should that make a difference?

Problem 9-L. Taking a Viewpoint

Adopting the perspective of the EPA, the companies with legal and financial obligations at the site, a group of activists from the affected community, or the local city government, choose the disposal method that you think is most appropriate. Argue as persuasively as you can within the context of the uncertainties of the analysis, making explicit the data that you feel are most important, as well as underlying value judgments. You may introduce additional relevant issues, and should point out the need for additional information that would be required to strengthen (or weaken) your argument.

Problem 9-M. Event-Decision Tree

Create an event-decision tree for this scenario. Discuss how it might benefit the various parties above. Can any of the three options be eliminated from all perspectives? For which outcomes do you expect the valuations to be most contested? Least contested? Why?

Problem 9-N. Regulatory Impact Analysis: Where the Rubber Meets the Road

Many risk professionals and practitioners feel that the real meat of the process of implementing health, environmental, technological, or financial risk assessments is tragically neglected, underappreciated, or at least misunderstood. The process of translating scientific discoveries, such as the risk to the ozone layer from continued chlorinated fluoro-carbon usage (Benedick 1991), is often far more transparent and linear than the process of crafting and then implementing legislation that provides genuine improvements in health and environmental conditions. Problem 9-M provides a forum to explore the process where "the rubber meets the road."

Bills written to implement risk policy are products of committee and compromise, where single words convey precise but wildly different meanings to Democrat and Republican elected officials, to lawyers, and to implementing agency officials. They are then further defined by the courts, often in prolonged trials and oversights. For example, the statutes, laws, acts, and executive orders that shape the actions of the U.S. Environmental Protection Agency, Food and Drug Admistration, Department of Immigration, Department of Agriculture, National Institute of Health, Center for Disease Control, and Department of Energy take a variety of forms. The Delaney Amendment to the Pure Food and Drug Act (1958), rescinded in 1997, called for the complete elimination of carcinogens from food additives, while the Federal Insecticide, Fungicide and Rodenticide Act (FIFRA) calls for a balancing of costs and benefits. The range of possible interpretations is daunting, but only opens Pandora's box for the wording in more recent bills that call for public protection from known or suspected harmful compounds and processes. Some of the wordings of risk-reducing acts include

- "Best available technology economically achievable (Clean Water Act)
- No "unreasonable adverse effects (Environmental Pesticide Control Act)
- "To the extent feasible" (Occupational Safety and Health Act)
- "To provide an ample margin of safety" (Clean Air Act)
- Products should not create an "unreasonable risk of injury" (Consumer Product Safety Act)
- "To the extent feasible (taking costs into account)" (Safe Drinking Water Act)
- Toxic wastes should constitute no "significant" hazard (Resource Conservation and Recovery Act)
- Highly toxic substances should be banned (Hazardous Substances Act)

Whew! This range of language, sometimes contradictory and invariably vague, contributes to a legal and regulatory quagmire. The language is created through negotiation in committees and amendments. In most cases, the meaning of such terms as "significant" is hammered out over years of hearings, meetings, and lawsuits. And while there is often talk about rationalizing the process, the eventual impact of a new regulation is never certain, and often differs considerably as applied than any of its many authors originally intended.

So what does each of these wordings mean in terms of quantifiable standards for contaminants in air, soil, and water, and what level of protection does each of these acts provide to children, adults, the average or maximum-exposed individual? Finally, how legal and enforceable is each of these laws? Will they be challenged by industry or environmental groups, each of whom knows that to tie up a bill in litigation for years, even if only to lose, often constitutes victory. Ralph Nader vowed to potentially lose every battle, but win the war. In this contentious environment, it is not too surprising (although perhaps troubling) that lawyers constitute the largest single contingent within the U.S. EPA (table 9-4; Morgenstern 1997).

How can one get a feel for and an introduction to this process? Short of interning at the U.S. EPA, FDA, or other agency, the best introduc-

Table 9-4 U.S. EPA Employees with Graduate Degrees: Ten Most Common Categories

Discipline	Ph.D.	M.A./J.D.	Total	Total %
Law	50	1,201	1,251	18.1
Biological science/life science	500	624	1,124	16.2
Engineering	104	992	1,096	15.8
Physical science	323	613	936	13.5
Business management/administration	6	463	469	6.8
Public administration	9	442	451	6.5
Social sciences and history	52	294	346	5.0
Health professional	31	157	188	2.7
conservation/renewable resources	23	154	177	2.6
Economics	31	85	116	1.7
U.S. EPA totals (all disciplines)	1,248	5,674	6,922	100%

tion involves direct immersion in the exact wording and implications of the law.

As the introduction above made clear, risk regulations are constantly evolving. Understanding requires a plunge. To solve this problem, you will need to access several recent pieces of risk regulation from the *Federal Register*.

From the perspective of the groups listed below, evaluate and then write a memo supporting or attacking the Risk Regulation Bill (S. 981). Two statements supporting the bill by the sponsors Senators Thompson (R-Tennessee) and Levin (D-Illinois) (available in the *Federal Register*) and one by the Natural Resources Defense Council (NRDC) opposing the bill are provided below. Pay particular attention to the implications of the Thompson/Levin bill and its predecessor the Glenn/Chafee bill for cost-benefit analysis to be used in evaluating new and existing regulations, and for the role of experts in a peer review process. You will need to consult the 1995 Glenn/Chafee bill and most probably the 1994 HR9 legislation.

Write memos from the positions of

i. The administrator's office of the U.S. EPA
ii. The administrator's office of the U.S. FDA
iii. The U.S. Federal Court of Appeals
iv. A senator from Delaware
v. A senator from California
vi. The Sierra Club
vii. General Motors
viii. A pulp and paper mill in Spokane, Washington
ix. The mayor's office, Trenton, New Jersey
x. The mayor's office, Eugene, Oregon
xi. The Union of Farm Workers of America
xii. The Everglades County Chapter of Friends of the Earth

STATEMENT OF SENATOR FRED THOMPSON
JUNE 27, 1997

Regulatory Improvement and Government Accountability

Mr. Thompson. Mr. President, I am pleased to be able to join with Senator Levin and several of our colleagues in introducing legislation to improve how the federal government regulates. This legislation is an effort by some of us to devise a common solution to the problems of our regulatory system. We have some real political differences among us, but we all share the same goals: clean air and water, injury-free workplaces, safe transportation systems, to name a few of the good things that can come from regulation. We also all share the goal of avoiding regulation which unnecessarily interferes in people's lives and businesses, which costs more than it benefits, or which—inadvertently—causes actual harm.

I am pleased we are introducing this bill with Senators Glenn, Abraham, Robb, Roth, Rockefeller and Stevens. They have all toiled in the fields to improve regulation.

It was in this spirit that the legislation we are introducing today was drafted. The Regulatory Improvement Act will promote the public's right to know how and why agencies regulate, improve the quality of government decision making, and increase government accountability and responsiveness to the people it serves.

The problem is that agencies sometimes lose sight of common sense as they create regulations. Then even well-intentioned rules can produce disappointing results.

Consider the airbag issue that has been in the news lately. The National Highway Transportation Safety Administration required high-force air bags to maximize the odds of survival for adult males in highway crashes. But the deployment force from these air bags can be so severe that they can injure children, women, and the elderly. Senator Kempthorne has spoken about the tragic death of a young girl from Idaho who was decapitated when an air bag deployed during a low-impact collision. The agency is now considering the use of an air bag cut-off switch to avoid these tragedies. But Mr. President, tragedies like this never should have occurred. We could have avoided needless deaths and injuries if the agency had carefully considered the risks that high-impact airbags pose to certain populations. I hope today's proposal will correct mistakes like this before they occur.

A second example is the removal of asbestos from our schools and other public buildings. Early in the 1980s, government scientists argued that asbestos exposure could cause thousands of deaths. Congress responded by passing a sweeping law that led cities and states to spend nearly $20 billion to remove asbestos from public buildings. After further research, EPA officials eventually concluded that ripping out the asbestos had been an expensive mistake. Ironically, removing the asbestos actually raised the risk to the public—because asbestos fibers become airborne during removal. This mistake never would have occurred if these increased risks had been considered in the first place. I hope that would change under the Regulatory Improvement Act.

Finally, let me mention our Superfund requirements. Superfund was passed with the good intention of cleaning up America's toxic waste sites. Unfortunately, things are not working as well as intended. Superfund has become a legal and regulatory maze where a good 90% of insurers' costs and 20% of liable parties' costs are spent on lawyers and consultants—not on cleaning up the environment. We also have to ask if we are focusing on the most important priorities. For example, Superfund imposes extremely stringent standards for cleaning up lead in groundwater. Now, this is a good rule in many cases, because lead can be very toxic to children. The problem is that we may be overlooking more direct threats to children from lead. For example, lead paint in old houses can be a greater threat to children's health than lead that may be under some industrial site where there are no children. Last congress, our Committee heard testimony about how the Superfund law requires groundwater in a Newark rail-yard to be cleaner than drinking water—at enormous cost. Now, if land is going to be used for industrial purposes, and no children will be there, does this make sense? The answer may be no—those requirements may not improve the environment much, but they may drive businesses out of Newark. Nobody wants to open a business near a Superfund site and risk being sued. No wonder our inner cities are starved for jobs. In the end, we may be hurting the very people we should be concerned about—the inner-city poor, those who already have to live with many risks in their daily lives, those who do not have clout here in Washington.

Virtually every serious student of the regulatory process agrees we can do better. One study by the Harvard Center for Risk Analysis found that if agencies simply set their priorities in a smarter way, we could save an additional 60,000 lives per year at no additional cost. Mr. President, we don't have a moment to lose when we could save more lives. We can set aside partisan politics, and we all can agree this is the right thing to do.

Since I became Chairman of the Governmental Affairs Committee, I have been working closely with Senator Levin to forge bipartisan legislation with three major purposes:

First, to promote the public's right to know how and why agencies make regulatory decisions. This legislation helps the public to understand agency decisions by directing agencies to—

Allow the public to comment and participate as rules are developed;

Disclose the benefits and burdens of major rules;

Disclose any environmental, health and safety risks a rule is designed to reduce, make those risks understandable by comparing them with other risks familiar to the public; and

Identify major assumptions and uncertainties considered in creating rules.

Second, to improve the quality of government decision-making. Careful thought, grounded in science, will help us to target problems and to find better solutions. We must carefully craft new rules to be effective and efficient. Agencies will carefully consider the benefits and burdens of rules and use good scientific and technical information. Agencies will seek out smarter ways to regulate, including flexible approaches such as outcome-oriented performance standards and market mechanisms. We must modernize and improve rules already on the books. Independent committees will advise agencies how to revise rules to substantially increase the benefits to the public.

And finally, to increase government accountability to the people it serves. The Act will require agencies to

Clearly present regulatory proposals so the public, the Congress and the President can understand the problem at hand and help find a solution;

Explain any legal impediment or other factor hindering the agency from issuing cost effective and sensible regulations, and describe any superior alternatives;

Disclose realistic estimates of any risks addressed;

Document changes made to proposed rules when the rules are reviewed by the Office of Management and Budget ("OMB");

Disclose contacts from persons outside the Executive Branch with OMB when it is reviewing proposed rules, since such contacts may represent outside influence.

Mr. President, while it is important to review what this legislation will accomplish, it also is important to note that this proposal avoids the contentious issues that thwarted agreement on legislation last Congress.

First, this legislation does not contain a "supermandate." That is, while we believe that cost-benefit analysis is an important tool to inform agency decision making, the results of the cost-benefit analysis do not trump existing law. The bill explicitly recognizes that sometimes an agency will issue a rule that would not pass a cost-benefit test. We only ask the agency to explain why it selected such a rule, including any legal impediment that hindered the agency from issuing a cost-justified rule.

Second, this bill does not contain a petition process that would allow outside parties to sue agencies in court to change particular rules that the litigant does not like. While we believe there are fruitful opportunities to update and improve old rules, we do not want to set up a review process that could create a litigation morass. Instead of a petition process, agencies will use independent advisory committees that would recommend a list of rules that could be improved to substantially increase net benefits to the public. The agency would defer to the recommendations of the advisory committee, but they could not be dragged into court if someone wanted a different rule to be reviewed.

Finally, this bill strikes a balanced approach to judicial review. We allow limited judicial review under the deferential arbitrary and capricious standard to ensure that agencies issue reasonable regulations using the tools of cost-benefit analysis and risk assessment. But this legislation does not provide a series of trip wires that could hinder agencies from performing their missions. In other words we realize the agencies may not be perfect in complying with this law. They may make mistakes from time to time. We won't imperil important regulations because the agency made honest mistakes. We just ask the agency to make reasonable and honest decisions, and the public deserves no less.

Mr. President, we are devoting vast resources to achieve our regulatory goals. By some estimates, the annual regulatory burden is nearly $700 billion per year—almost $7,000 for the average American household. Our regulatory goals are too important, and our resources are too precious, to spend this money unwisely.

The Regulatory Improvement Act will ensure that agencies conduct better economic and scientific analysis before they issue regulations. Government will be more open to the public, will better explain the problem, and will consider the best available information to solve the problem. Agencies will consider the benefits and burdens of different regulatory alternatives so we can reach the most sensible solutions. And agencies will modernize old rules on the books to increase the benefits to the public. In the process, we won't sacrifice our important national goals and values. We can make our government more effective, more open, and more accountable than ever.

FLOOR STATEMENT OF SENATOR LEVIN
ON REGULATORY IMPROVEMENT ACT

Mr. President, today Senator Thompson and I are joined by Senators Glenn, Abraham, Robb, Roth, Rockefeller and Stevens in introducing the Regulatory Improvement Act of 1997. The bill would put into law (1) basic requirements for cost-benefit analysis and risk assessment of major rules, (2) a process for the review of existing rules where there is a possibility of achieving significantly greater net benefits, and (3) executive oversight of the rulemaking process. It builds on the bipartisan Roth-Glenn bill that was unanimously reported out of the Governmental Affairs Committee in 1995.

This bill would require agencies, when issuing rules that have a major impact on the economy or a sector of the economy, to do a cost-benefit analysis to determine whether the benefits of the rule justify its costs and to determine whether the regulatory option chosen by the agency is more cost effective or provides greater net benefits than other regulatory options considered by the agency. If the rule involves a risk to health, safety or the environment, the bill requires the agency to do a risk assessment as part of the analysis of the benefit of the rule.

The bill also requires agencies that issue major rules to establish advisory committees to identify existing rules that the agency should consider for review because they have the potential, if modified, to achieve significantly greater net benefits. It would also codify the review procedure now conducted by the Office of Information and Regulatory Affairs (OIRA) and require public disclosure of OIRA's review process.

The bill is significantly different from S. 343, the Dole-Johnston bill which I strongly opposed and which was rejected by the Senate in the 104th Congress.

—It does not create a "supermandate" that would amend existing laws nor does it contain mandatory "decisional criteria" that would establish new standards for an agency to meet. It does require agencies to conduct cost-benefit analyses for major rules and explain whether the benefits of the rules justify the costs and whether the rule is more cost-effective than the other alternatives considered by the agency. It does not mandate the outcome of the process, only the process itself.

—It does not provide for judicial review of the process for, or the contents of, the cost-benefit analysis or risk assessment. The cost-benefit analysis and risk assessment are made part of the rulemaking record for judicial review of whether the final rule is reasonable.

—It does not provide for a petition process for challenging existing rules. It provides for advisory committees to identify rules for possible review, gives the agency head the discretion to select rules for review especially taking into account the resources of the agency, and requires the agency to review the rules scheduled for review in 5 years.

Mr. President, many people think that when many of us fought hard against the Dole-Johnston bill that we didn't really want to reform the regulatory process. Well they are wrong. Many of us were disappointed that we were unable to pass a comprehensive regulatory reform bill in the last Congress. We weren't going to support bad reform, but that doesn't mean we didn't want to see good reform. Those of us who believe in the benefits of regulation to protect health and safety have a particular responsibility to make sure that regulations are sensible and cost-effective. When they aren't, the regulatory process—which is so vital to our health and well being—comes under constant attack. By providing a common sense, moderate and open regulatory process, we are contributing to the well being of that process and immunizing it from the attacks on excesses.

Mr. President, I've fought for regulatory reform since 1979, the year I came to the Senate. I even had as part of my platform back in 1978, the legislative veto—which would give Congress the chance to block excessively costly and burdensome regulations before they take effect. That was my battle cry for years. I worked with former Senator Boren, for instance, trying to get an across-the-board legislative veto bill enacted into law. Last Congress we were finally able to get a version of that adopted.

I was also the author of the Regulatory Negotiation Act which was passed in 1990 and reauthorized in 1995 to encourage agencies to use the collegial process of negotiation in developing certain rules in order to avoid the delays and costs inherent in the otherwise adversarial process.

As for an overall regulatory reform bill, I've supported such legislation since 1980, when the Senate first passed S. 1080, the Laxalt–Leahy bill, only to have it die later that year in the House.

At the same time, I took a strong stand against several damaging regulatory reform proposals from the House including an overall moratorium of regulations and against the Dole–Johnston bill in the Senate. I will not support any regulatory reform proposal that I believe would roll back important environmental, public health and safety protections. Nor will I support any regulatory reform proposal that I believe will lead to gridlock in the agencies or the courts. We certainly don't need that.

We do need better cost-benefit analysis and risk assessment, more flexibility for the regulated industries to reach legislative goals in a variety of ways, more cooperative efforts between government and industry and less "us versus them" attitudes.

Based on these common principles, Senator Thompson and I have been working for months on this legislative proposal that I hope will yield a more rational and fair regulatory process and better, more flexible, more cost effective and more enforceable regulations.

Let me highlight some important features of this legislation.

First, we say right from the beginning, in the section on findings, that cost-benefit analysis and risk assessment are useful tools to help agencies issue reasonable regulations. But they are only tools; they are not the sole basis upon which regulations should be developed or issued. They do not, we explicitly state, they do not replace the need for good judgment and the agencies' consideration of social values in deciding when and how to regulate.

We define benefits very broadly—expressly taking into account nonquantifiable benefits. There is nothing in this bill that suggests that the assessment of benefits by an agency should be only quantifiable. On the contrary, this bill explicitly recognizes that many important benefits may be nonquantifiable, and that agencies have the right and authority to fully consider such benefits when doing the cost-benefit analysis and when determining whether the benefits justify the costs. We emphatically do not intend for the benefits part of the equation in the cost-benefit analysis to be limited to merely those benefits that are quantifiable.

We direct the agencies to consider regulatory options that provide flexibility, where possible, to the regulated parties. I have been a longtime proponent of performance standards in regulations and not the so-called "command and control" approach. This bill urges the agencies to include in its identification of possible regulatory approaches those that permit flexibility in achieving the required goal, either through performance standards or market type mechanisms.

The definition of major rule, to which the provisions of this bill apply, is limited to those with a $100 million impact on the economy and those otherwise designated by the Administrator of the Office of Information and Regulatory Affairs (OIRA).

The bill requires an agency issuing a major rule to evaluate the benefits and costs of a "reasonable number of reasonable alternatives reflecting the range of regulatory options that would achieve the objective of the statute as addressed by the rulemaking." I am quoting these words, because they are significant. The bill doesn't require an agency to look at all the possible alternatives, just a reasonable number; but it does require the agency to pick a selection of options that are available to it within the range of the rulemaking objective.

This cost-benefit analysis, of which any risk assessment would be a part, is intended to be transparent to the public—that is, those of us outside the agency (Congress, the regulated community, the beneficiaries of the regulation, the general public) should be able to see and understand the thinking the agency used to select the regulatory option it did, as well as the underlying scientific and/or economic data. Agencies should not hide the important information that forms the basis of their regulatory actions.

Another important provision of this bill is the one that requires the agency to make a reasonable determination whether the benefits of the rule justify the costs and whether the regulatory option selected by the agency is substantially likely to achieve the objective of the rulemaking in a more cost effective manner or with greater net benefits than the other regulatory options considered by the agency. This is not in any way a decisional criteria that the agency must meet. This only requires the agency to make its assessment. And, if, as the agency is free to do, it chooses a regulatory option where the benefits do not justify the costs or that is not more cost effective or does not provide greater net benefits than the other options, the agency is required to explain why it did what it did and list the factors that caused it to so. Those factors could be a statute, a policy judgment, uncertainties in the data and the like. There is no added judicial scrutiny of a rule provided for or intended by this section. The final rule must still stand or fall based on whether the court finds that the rule is arbitrary or capricious in light of the whole rulemaking record. That is the current standard of judicial review.

The bill says that if an agency "cannot" make the determinations required by the bill, it has to say why it can't. Use of the word "cannot" does not mean that an agency rule can be overturned by a court for its failure to pick an option that would permit the agency to make the determinations required by the bill. The agency is free to use its discretion to regulate under the substantive statute, and there is no implication that such rule must meet the standards described in the determinations subsection. It does mean, though, that the agency is required to make such determinations and let the public know why it picked the regulatory option that it did, and if it can't say, or "determine," that the regulatory option it chose is the most cost effective or provides greatest net benefits, it must say why it chose it. This legislation requires only that the agency be up front with the public as to just how cost-beneficial and cost effective its regulatory proposal is.

The risk assessment requirement in this bill, unlike previous bills, is not unduly proscriptive. It establishes basic elements for performing risk assessments, many of which, again, will provide transparency for an agency's development of a rule, and it requires guidelines for such assessments to be issued by OIRA in consultation with the Office of Science and Technology Policy.

Peer review, Mr. President, is required by this bill for both cost-benefit analyses and risk assessments, but only once per rule. Peer review is not required at both the proposed and final rule stages. There is great concern in the public interest community, that there will not be sufficient personnel available with appropriate expertise and independence to serve on each of these peer review bodies. I am hoping to pursue that issue at greater length during our committee hearings.

There is a similar concern by the public interest sector as to the availability of a balanced cross-section of individuals to serve on the advisory committees required for the review of rules. Service on such bodies obviously takes time and expertise and both of those cost money. I hope we can also address the concerns about the possibility of inadequate levels of participation by groups and interests which have fiscal constraints that could preclude their full participation.

Mr. President, the review of rules provision in this bill is also a reasonable approach. Unlike past proposals, it does not provide for an automatic sunset of a rule that is not reviewed pursuant to the schedule. Rather it provides for the agency to determine during the review period of rules it chooses to review whether it is going to continue, modify or repeal the rule under review. If it fails to make that determination and take the appropriate action, the agency can be sued under the existing provision of the Administrative Procedure Act to force agency action unlawfully withheld.

Rules would be scheduled for review under the provisions of this bill, only at the discretion of the agency head. However, the public would know the list of rules recommended for review by the advisory committee. The advisory committee would recommend those rules for review that, if modified, could result in substantially greater net benefits to society. That is the standard the committees are supposed to apply. The agency must review the recommendations of the advisory committee and develop a schedule for review of rules taking into account the resources available to the agency to conduct such a reviews.

Judicial review has been of great concern to those of us who want real regulatory reform without bottling up important regulations in the courts. There is no judicial review permitted of the cost-benefit analysis or risk assessment required by this bill outside of judicial review of the final rule. The analysis and assessment are included in the rulemaking record, but there is no judicial review of the content of those items or the procedural steps followed or not followed by the agency in the development the analysis or assessment. Only the total failure to actually do the cost-benefit analysis or risk assessment would allow the court to remand the rule to the agency.

Finally, Mr. President, the bill puts into law the requirement that the President establish a process for reviewing rules and coordinating federal agency regulatory actions. Despite over 15 years of Executive Orders that impose such a requirement, Congress has yet to put such a responsibility of the President into law. This bill would do that. And with that responsibility goes the obligation of the President, acting through OIRA, to make public the process and results of its review of agency rules. This is an important element of accountability, and such disclosure should not depend upon the whim of the President but rather on the requirements imposed by permanent law.

So those are some highlights. Senator Thompson has committed to hearings on the bill. Everybody will be given an opportunity to comment and identify potential problems and possible improvements.

I believe this bill will improve the regulatory process, will build confidence in the regulatory programs that are so important to this society's well-being, and will result in a better—and I believe—a less contentious regulatory process.

Thank you, Mr. President, and I yield the floor.

NRDC PRESS RELEASE

Eight Crucial Ways that the New Regulatory Reform Bill S. 981, Undermines Vital Health, Safety and Environmental Protections

While sponsors have indicated that this is not the intent, as crafted, S. 991 would undermine crucially important health, safety, and environmental safeguards, including food safety and drinking water protections enacted only last Congress. The bill's singular emphasis on inflexible economic estimates will weaken current laws calling for protective rulemakings, empower courts to overturn important safeguards, and dramatically tilt the regulatory playing field in favor of big businesses able to deploy legions of economic consultants and lawyers to challenge public protections. As explained below, in a number of crucial respects S. 991 is a far greater threat to public health protections than the Glenn/Chafee regulatory reform proposal, offered as a substitute for overtly destructive regulatory reform legislation in the last Congress.

1) The so-called "net benefits" test in S. 981 would require agencies to translate the various benefits of safeguards—from healthier children to safer foods—into a dollars and cents total that pushes agencies toward less protective rules. Section 623(c)(3) provides that, in addition to satisfying a broad cost-benefit requirement, agencies charged with health, safety or environmental protection must also determine if now safeguards meet quantitative cost-effectiveness or net benefits tests. While this may sound reasonable, this insidious test presumes that in every instance there is a single approach to protecting health, safety or the environment which can be mathematically demonstrated to have the greatest "net benefit" or be the most cost-effective. The reality of course is that policy choices often reflect priorities for health or environmental protection which do not readily translate themselves into such neat numerical frameworks. Even so, agencies will be required to rely on economic assumptions to generate precise dollar estimates for the benefits of regulation, including for example the dollar value of a human life, a healthy child, an intact ozone layer or a clean lake. (Last Congress' Glenn/Chafee bill did not include any comparable "net benefits" test.)

For example, in developing a rate to prevent the sale of unsafe cribs under S. 981, the Consumer Product Safety Commission would under section 623(c)(3) have to estimate the number of infant lives saved, decide how many dollars each infant's life is "worth," estimate the cost of designing safer cribs and then calculate the "net benefit" for a variety of regulatory options—ranging from doing nothing to doing the most the agency is authorized to do. The "net benefits" would be the dollar value of saved infants lives minus the *costs* of safer cribs. Under section 623(c)(3)(B) the agency is to select the option with *the* greatest "net benefits" unless it "cannot reasonably" do so.

2) S. 981 includes no "savings clause" to assure that existing health, safety, and environmental laws remain intact. Without a provision expressly preserving existing laws, the various requirements of S. 981 are likely to be interpreted to override existing legal protections. Assertions that this bill does not include a "supermandate" that overrides existing laws are therefore not accurate. This is especially true with regard to the inflexible "net benefits" test and associated requirement in Section 623 (c)(3). (Last Congress' Glenn/Chafee bill did include a savings clause expressly preserving existing laws.)

3) Agencies are discouraged from making health, safety or environmental protection a priority. A government agency seeking to issue a more protective rule that fails to pass the bill's tough quantitative tests, can go forward only if the Agency determines that it "cannot reasonably" select a rule allowing it to make the net benefit or cost-effectiveness determinations. (Section 623(c)(3)(B).) "Cannot reasonably" is a sweeping and undefined term that appears to impose a nearly impossible burden. (Last Congress' Glenn/Chafee bill did not impose any such burden. Instead agencies were simply instructed to explain whether the required cost benefit determination could be made.)

4) The judicial review provision in S. 981 appears to empower courts to overturn as "arbitrary and capricious" any rule where the agency has not met the economic tests called for in the bill. Section 627(d) expressly provides that the court "shall consider" the cost-benefit and risk assessment determinations required by the bill in determining whether the final rule is "arbitrary or capricious" and should therefore be overturned. Courts are allowed to disregard this directive only if the agency can demonstrate that such findings are not "material to the outcome of the rule." (Last Congress' Glenn/Chafee bill *did not* include any such directive for reviewing courts. That bill simply provided for the regulatory analyses to be "part of the whole record" available to the court.)

5) Unlike traditional objective "peer review" panels, the "peer review" panels established in S. 981 to review risk assessments and cost benefit analyses allow the participation of individuals with a direct conflict of interest. Section 625(b)(1)(B) would allow participation by individuals with a direct financial interest in the rulemaking, including individuals affiliated with parties subject to regulation under the rule, provided only that such financial interest is disclosed. This fact, in combination with the huge financial and manpower advantages of the private sector over the public interest community, suggests that the peer review panels called upon to oversee regulatory analyses are likely to be heavily weighted toward experts with industry affiliations. The concept of allowing individuals with a direct financial interest

to critique the analytical foundation for new rules, which they can later go on to challenge in court, is a grotesque distortion of the basic fairness that current law, in particular the Administrative Procedure Act, has long sought to assure in the regulatory process.

6) The Advisory Committee established under S. 981 is comprised not of independent experts but of parties "affected by the regulations," (section 632(b)(1)) and can, once again, include individuals with a direct financial conflict of interest. This Committee is charged with the crucial task of identifying existing rules which will be candidate for review and revision under the agency "lookback process." Section 632, which provides authority for the advisory Committee, fails to preclude or otherwise address the participation of individuals with a financial interest directly affected by rules under review, or to even require disclosure of such conflicts. (Last Congress' Glenn/Chafee bill did not include any comparable advisory committee. It did, however, have an extremely problematic petition process to provide the same function.)

Even safeguards already on the books would be subject to challenge. Agencies would select existing rules to review based on the list compiled by a new Advisory Committee on Regulation. The agency would, once again, be required to generate dollar value estimates for the health and environmental benefits of the rule, and then determine whether the rule can be revised to increase net benefits. Should the agency decide to keep the safeguard in place after the "lookback" study, section 633(1) of the bill empowers courts to overturn that decision, and send the rule back to the agency, if polluters, tobacco companies, or other regulated parties can show that another alternative would "substantially increase net benefits." This provision is an open invitation for big businesses, armed with massive teams of economic consultants and lawyers, to challenge and weaken popular existing protections ranging from cigarette labeling to clean water standards.

8) Although authors of S. 981 have indicated that they do not intend to allow judicial review of the process for, or contents of, the various regulatory analyses called in the bill the legislation does not include a provision specifically barring such review. Court review of such matters would almost certainly tie up the agency in lengthy and pointless litigation, that would serve mainly to delay public health protection. (Last Congress' Glenn/Chafee bill did include a provision specifically barring such judicial review.)

There is no record supporting the need for this sweeping revision of the current system for protecting our health, safety and environment. Government agencies are already fully analyzing the costs and benefits of their actions under both executive orders. and several significant pieces of recently enacted legislation, including the Small Business Regulatory Enforcement Fairness Act, the Unfunded Mandates Reform Act, amendments to the

Paperwork Reduction Act, and the Regulatory Accounting rider to the 1997 Omnibus Appropriations law. These laws should be given an opportunity to work, rather than overwhelmed with a series of broad new bureaucratic mandates.

References

Benedick, R. E. (1991). *Ozone Diplomacy.* Cambridge, MA: Harvard University Press.

Dasgupta, P. and Mäler, K.-G. (1994). Poverty, Institutions, and the Environmental-Resource Base." World Bank environment paper no. 9. Washington, DC: The World Bank.

Kleindorfer, P. R., Kunreuther, H. C., and Schoemaker, P. J. H. (1993), *Decision Sciences: An Integrative Perspective.* New York, NY: Cambridge University Press.

Morgan, M. G. and Henrion, M. (1990). *Uncertainty: A Guide to Dealing with Uncertainty in Quantitative Risk and Policy Analysis* New York, NY: Cambridge University Press.

Morgenstern, R. D. (1997). "The legal and institutional setting for economic analysis at EPA." In Morgenstern, R. D. (Ed.) p. 15. *Economic Analyses at EPA: Assessing Regulatory Impact*, Washington, DC: Resources for the Future.

Tengs, T. O., Adams, M. E., Pliskin, J. S., Safran, D. G., Siegel, J. E., Weinstein, M. C., and Graham, J. D. (1995). "Five-Hundred Life-Saving Interventions and Their Cost-Effectiveness," *Risk Analysis* 15(3):369–90.

10

Risk Perception and Communication

Introduction

We began this book with a quote from Harry Otway (1992) that summarized his evolution in thinking about and performing and evaluating risk assessments. He reflected that during his career:

> I began by trying to quantify technical risks, thinking that if they were "put into perspective" through comparison with familiar risks we could better judge their social acceptability. I am ashamed now of my naiveté, although I have the excuse that this was more than twenty years ago, while some people are still doing it today.

The intellectual realization outlined by Otway is also the basis of a common refrain by researchers and practitioners involved in risk analysis who note that the "technical" work is in the end the easy part. It is the process of translating analysis to policy-relevant information, advice, and action for decision making that is truly challenging. Information (about risk or otherwise) is of little value if not communicated to others. Learning how to present and communicate information not only provides the initial step in making risk analysis relevant, it is also critical in helping you to think through what calculations you are doing, and why, and for whom you are doing them.

Baruch Fischhoff, who has been involved in risk communication for several decades, has arrived at a conclusion that parallels Otway's. He summarizes his experience in table 10-1. In this chapter, which uses lessons from the rest of the book to examine risk perception and risk communication, we will see that the "all of the above" attitude is often the best policy when considering how analysis is likely to be interpreted and used. The best way to approach the risk communication problems in this chapter is to consider not what you want to say, but what message will be heard. Part of the goal of this chapter is to move the reader to identify and analyze not simply what information you want to give others, but what information others want and need.

Table 10-1 Developmental Stages in Risk Management
(Ontogeny Recapitulates Phylogeny)

All we have to do is get the numbers right
All we have to do is tell them the numbers
All we have to do is explain what we mean by the numbers
All we have to do is show them that they've accepted similar risks in the past
All we have to do is show them that it's a good deal for them
All we have to do is treat them nicely
All we have to do is make them partners
All of the above

Source: Fischhoff 1995.

Problem 10-1. Same Numbers, Different Stories

Suppose that you represent a regulatory agency reviewing an application for a new facility, located in a city of 500,000 people. The owners have provided information indicating that, using conservative human health calculations, the facility will cause an additional annual individual risk of 7.2×10^{-7} premature fatalities (from all possible deaths, including cancers, accidents, etc.), which is below the 10^{-6} standard set by your agency. A community organizer uses the company's own numbers to show that, in fact, the risk is almost entirely incident upon a small low-income community of 2000 near the facility. In addition, she points out that your agency has approved thirteen similar facilities in the area over the past fifteen years, and a total of six in the rest of the city, and concludes that your agency has approved the premature deaths of 5% of the members of her community over the next twenty years. The company threatens to sue if the project is not approved; the community group threatens the same if it is approved.

    a. Calculate (i) the annual risk to the average city resident from all sources; (ii) the annual risk of premature fatalities in the city from all sources; (iii) the annual risk to the average community resident from the new facility; (iv) the annual risk to the average community resident from the thirteen existing facilities; (v) the annual risk of premature fatalities to the community from all sources; and (vi)

total annual deaths from the thirteen existing and one new facilities over a twenty-year period.
b. What is an "average" resident?
c. Is the company estimate right? The community organizer's estimate? Discuss.

Does any of the estimates in (a) "best" represent the situation?

Solution 10-1

In working this problem, begin by considering the degree to which risk information is not value neutral. The fact that a study or calculation was performed at all suggests to many that there is a risk (regardless of the result obtained). Any report or analysis will mean different things to different people.

Solution 10-1a(i)

Assuming that each poses a 7.2×10^{-7} risk,

$$\begin{aligned}
\text{risk} &= \text{number of facilities} \\
&\quad \times \text{impact of each facility per person per year} \\
&= 19 \text{ facilities} \times 7.2 \times 10^{-7} \text{ fatalities/facility/year/person} \\
&= 1.4 \times 10^{-5} \text{ fatalities/year/person}.
\end{aligned}$$

Solution 10-1a(ii)

$$\begin{aligned}
\text{risk} &= \text{number of facilities} \\
&\quad \times \text{impact of each facility per person per year} \\
&\quad \times \text{number of people exposed} \\
&= 19 \text{ facilities} \times 7.2 \times 10^{-7} \text{ fatalities/facility/year/person} \\
&\quad \times 500,000 \text{ people} \\
&= 6.8 \text{ fatalities per year}.
\end{aligned}$$

Solution 10-1a(iii)

If the risk is entirely incident on the small community, then

$$\text{risk} = (7.2 \times 10^{-7} \text{ fatalities/city resident/year})$$
$$\times (500{,}000 \text{ city residents})/(2000 \text{ community residents})$$
$$= 1.8 \times 10^{-4} \text{ fatalities/community resident/year.}$$

Solution 10-1a(iv)

$$\text{risk} = 1.8 \times 10^{-4} \text{ fatalities/community resident/facility/year}$$
$$\times 13 \text{ facilities}$$
$$= 2.3 \times 10^{-3} \text{ fatalities/community resident/year.}$$

Solution 10-1a(v)

Assuming that the risk from the six more distant facilities is orders of magnitude lower than that from the local facilities,

$$\text{risk} = 2.3 \times 10^{-3} \text{ fatalities/community resident/year}$$
$$\times 2000 \text{ residents}$$
$$= 4.6 \text{ fatalities per year in the community.}$$

Solution 10-1a(vi)

$$\text{risk} = (1.8 \times 10^{-4} \text{ fatalities/community resident/facility/year})$$
$$\times (14 \text{ facilities}) \times (2000 \text{ residents}) \times (20 \text{ years})$$
$$= 100 \text{ premature fatalities in the community over the}$$
$$\text{next 20 years.}$$

Solution 10-1b

The "average" resident is clearly a problematic concept in this case, since the numerical average does not represent a real person. Rather, there is a bimodal distribution (or multimodal, depending on the locations of the other six facilities) of individuals, a low-risk group outside the community, and a high-risk group within it. Consequently, while one can calculate the risk to the average resident, it does not describe *any* real individual risks, which are either much higher or much lower, depending on location.

Solution 10-1c

This question highlights the importance and impact of defining the terms of analysis (Fischhoff et al. 1984). It is hard to argue that either set of numbers is "right" or "wrong," since all are numerically correct within the analytical framework. What differs dramatically is what the numbers mean to us when framed in different ways. Importantly, what if there were no community organizer, and it hadn't occurred to you as the assessor to think about incidence, just about aggregates? It is likely that the 7.2×10^{-7} risk, estimated using conservative calculations, would seem entirely acceptable. Yet, when posed as a group of decisions leading to a significant fraction of deaths, the decision appears entirely unacceptable.

The National Research Council (1996) recently addressed how risk characterization is most appropriately approached. The study concluded that the terms of analysis must be defined by a broad set of "interested and affected" parties, and that the needs of these parties must determine what analysis is done. Similarly, the process of characterizing risks should not rely only on analysis, but should be a marriage of analysis and deliberation. While achieving this ideal is not simple (and the NRC identified few successful cases), it should be the goal of any risk characterization effort.

Problem 10-A. Opening Dialogue

Suggest additional ways to get the community group and the company to start working together.

Problem 10-B. Explaining Numbers

Briefly explain in terms a reasonably intelligent person with a high-school education could easily understand the differences among the following:

a. A 10^{-6} individual risk of fatality

b. A risk of one fatality given an exposed population of one million
c. An individual risk of one in one hundred thousand
d. An aggregate risk of one in one hundred thousand

Problem 10-2. Framing a Question: Loss or Gain?

Problem 10-1 looked at how "framing" (or presenting) numbers can make them mean very different things. This problem looks instead at how different contexts (frames) can cause individuals to express very different preferences for what are in fact numerically identical options.

A well-understood process in individual decisions about risk is that people are "risk averse" (avoiding) when considering potential gains. This means that, in general, when people are deciding whether to make a bet, they will require better odds when the stakes are high. On the other hand, when they face potential losses, they are "loss averse." Knowing this would lead to a straightforward decision rule if the point of reference were well defined, allowing us to differentiate between gains and losses. This problem suggests that a single reference may not exist.

Assume you ask a large number of doctors to make a decision about a hypothetical serious disease outbreak for which two possible treatments exist. You present half the doctors with case A below, and the remaining with case B.

a. Discuss the numerical meaning of these two cases.
b. For those doctors presented with case A, would you expect more to opt for (i) or (ii)? Why? Would you expect more or fewer doctors to opt for (iii) over (i)? Why?
c. Discuss the implications of this problem in a decision-making context.

Both options are for a population of six hundred facing a disease that is certain to be fatal if left untreated.

Case A:

Option (i): A treatment that gives a 1/3 chance to save everyone, but a 2/3 chance of saving no one, or
Option (ii): A treatment that is certain to save 200.

Case B:

> Option (iii):A treatment that has a 2/3 chance of letting everyone die, but a 1/3 chance that no one will die, or
>
> Option (iv): A treatment that is certain to let 400 die.

Solution 10-2

Solution 10-2a

All four options have the same expected value (200 lives saved, 400 lives lost). However, (i) and (iii) are probabilistic (unknown outcome), while (ii) and (iv) are deterministic (known outcome).

Numerically, (i) and (iii) are identical probabilities, with a 1/3 chance of everyone living, and a 2/3 chance of everyone dying. Likewise, (ii) and (iv) are identical, with 200 people living and 400 dying.

The difference between the cases is that (A) is framed as potential gains (people are saved), and (B) as potential losses (people die).

Solution 10-2b

Studies suggest that the number who opt for (i) will be much greater than the number who opt for (ii). This is because most people prefer a sure gain than a gamble for the same expected gain. However, in case B, the "certain to die" option will drive the decision. Most people would rather avoid a sure loss if they can gamble for the same expected loss. Studies suggest that not only will more choose (iii) than (i), but the number choosing (iii) will be considerably greater than the number choosing (iv). This type of finding has been confirmed in a number of decision-making contexts (Kahneman and Tversky 1979).

Keep in mind, however, that studies of this type are artificial: they give the decision maker a limited context (only case A or B, and not both) and limited time to respond. Consequently, inferring how people make decisions on real issues from these studies requires as many assumptions as do the high-to-low dose extrapolation techniques presented earlier (Fischhoff 1991). One should not expect people given a more diverse set of information, or time to consider the alternatives carefully, to react as they do in experimental settings. One important lesson, however, is that looking at a problem from a variety of perspectives can reduce some of the potential for framing bias, thereby improving the decision process.

Solution 10-2c

While this is a straightforward example, it has profound implications for risk communication. Clearly, it is not simply the value of the numbers, nor the preferences of an individual, that influences decisions. Rather, whether risk information is framed as a loss or gain has a dramatic impact on the final choice. In addition, it is not only the inexperienced who exhibit this behavior, but people who are in positions to make this type of choice. Kahneman and Tversky (1979) first recognized this effect and have termed it "prospect theory," referring to the observation that we make decisions based on whether we perceive prospective gains or losses.

One positive lesson that can be drawn from this is that looking at or communicating information in a variety of ways, for example, giving both (A) and (B), may promote better decisions. More cynically, one should be alert for how others can frame information to promote one option over another.

Problem 10-C. A Project to "Restore" or "Improve" a Wetland

Based on the above information, how would you expect the following descriptions of a wetland area project to affect preferences? Discuss.

a. This project will invest public funds in the restoration of the Green River to its preindustrial state
b. This project will spend tax dollars to improve the quality of the Green River.

Problem 10-D. A Little Bit or a Lot? Violent Agreement on the Numbers

In an article promoting the siting of a high-level radioactive waste storage facility, Wolfe (1996) argues that "[i]f all the country's high level nuclear waste...were collected on a football field, it would be only 9 feet deep." He concludes that this is not a lot of material to store, so we might as well do it. In contrast, Flynn et al. (1989) are concerned about the "scope of the hazard," observing that by the year 2000, there will be enough high-level nuclear waste to pile fifteen feet deep on a football field, and that it would be rash to commit such large amounts to an uncertain technology.

    a. Discuss the two estimates of waste volume. Are they significantly different?

    b. Why do the two different authors use what are essentially the same numbers, and even the same framing, published several years apart in the same journal, yet expect readers to react in opposite ways?

    c. What do these two descriptions say about the role of analysis in the high level radioactive waste storage debate?

Problem 10-3. Risk Level, Risk Perceptions, and Psychometric Models

Over the past twenty years, a wide range of non-numerical attributes have been found to affect how likely we are to accept risks. This problem considers how several well-studied characteristics cause individuals to be more or less concerned about specific risks.

From each of the following pairs of potential hazards, which would you expect the people (on average) to worry about more? Why?

    a. (i) Accidents while you are driving a car, or (ii) accidents from riding in a car driven by someone you've never met

    b. (i) Risk from electromagnetic fields or (ii) risk from automobile accidents

    c. (i) Risk from caffeine, or (ii) risk from radioactive waste

    d. (i) A 10^{-5} lifetime risk of cancer due to well water contaminated with metals leached from an abandoned hazardous waste site, or (ii) A 10^{-5} lifetime risk of cancer due to well water contaminated by metals leached from a naturally occurring geologic formation

    e. (i) Risk from car accidents, or (ii) risk from airplane accidents

    f. (i) Risk from chemical X, a food preservative, or (ii) Risk from chemical Y, a food contaminant

Solution 10-3

Keep in mind that the solutions below are based on studies of many individuals. Since there is considerable variation from person to person, one must use caution when assessing how different risks might be perceived.

Solution 10-3a

Another characteristic of risks that influences how they are perceived is the degree to which one perceives them to be controlled. This would suggest that people will be more comfortable with their own driving than with that of a stranger.

Solution 10-3b

How well a particular risk is known, and how familiar people are with it, tend to influence the extent to which they worry about it. One would expect people to be more concerned about electromagnetic fields, because they are both unknown and ubiquitous. The issue of voluntary versus involuntary risks may also be important in comparing these two risks.

Solution 10-3c

People are more concerned about some risks that evoke feelings of "dread." "Cancer," "radioactivity," and "hazardous waste" are at the high end of the dread spectrum. Caffeine, on the other hand, is unlikely to be associated with dread. "Stigma" is a term that has recently been applied to a number of risks where numbers appear to be entirely irrelevant to perceptions. However, the use and meaning of "stigma" continue to be contentious (Kunreuther and Slovic forthcoming).

Solution 10-3d

Even when the risk numbers and causes are identical, people care more about man-made than naturally occurring risks.

Solution 10-3e

Although many more people are killed each year in car accidents than in airplane accidents, the latter are often of greater concern. A salient characteristic of airplane accidents is that they are catastrophic—when there is a major accident, many people die. In contrast, car accidents rarely involve the death of more than a few people.

Solution 10-3f

A risk perceived as having offsetting benefits may be of less concern than one with no perceived benefits. As a first cut, this appears sensible, but it might also constitute a form of "double counting," if both the benefit and the perception of that benefit are included in a decision process. It can even have a perverse effect if the perceived benefit does not exist.

Familiarity, dread, and control are well documented characteristics that people attribute to risks. Figure 10-1 is part of a "cognitive map" of risks along two "factors," or combinations of characteristics. Table 10-2 summarizes the characteristics comprising each "factor." (Both are from Slovic 1987.) Each of the pairs above involves more than one of these effects. What is important about such cognitive maps is not the diverse dimensions themselves, but what understanding the dimensions can give us about what is important when making risk decisions.

Problem 10-E. Cognitive Maps

Suggest where you think each of the fourteen risks in problem 10-3 would fit on the cognitive map in figure 10-1. List and explain several potentially influential characteristics in addition to those we have identified. Discuss how these characteristics affect people's attitudes toward the risks in each pair, and suggest how they might play into individual and policy decisions for

    a. Global warming
    b. Automobile risks

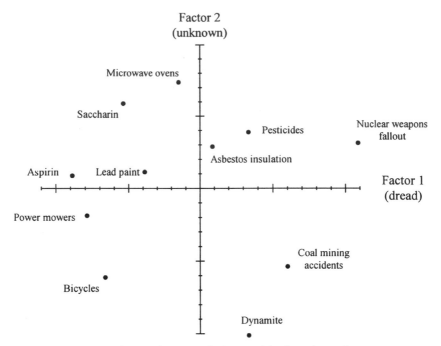

Figure 10-1. Psychometric map of eleven risks based on the extent to which they are considered "unknown" and "dread." After Slovic 1987.

   c. Installing high-voltage power lines
   d. Shipping hazardous waste from Zaire to Niger
   e. Shipping hazardous waste from the United Kingdom to Niger

Problem 10-F. Radiation By Any Other Name

For five years, John W. was tasked by New Jersey's governor with finding a community that would "host" a permanent disposal facility for low-level radioactive wastes. Mr. W. occasionally joked that he wished he could use the term "inertly challenged material" in lieu of "radioactive waste." Why might this have been an issue? Would such a name change have affected the chances of finding a disposal site? Why or why not?

Table 10-2 Components of Dread and Unknown Factors

Factor 1 (Dread)		Factor 2 (Unknown)	
Low	High	Low	High
controllable	uncontrollable	observable	not observable
not dread	dread	known to those exposed	unknown to those exposed
not global catastrophic	global catastrophic	effect immediate	effect delayed
consequence not fatal	consequences fatal	old risk	new risk
equitable	not equitable	risks known to science	risks unknown to science
individual	catastrophic		
low risk to future generations	high risk to future generations		
easily reduced	not easily reduced		
voluntary	involuntary		

Source: Slovic 1987.
Note: Factor analysis indicates a strong correlation among the items in each "factor." Consequently, they are grouped for this psychometric mapping.

Problem 10-G. Surfers and the Sun Revisited

Suppose that you have found that individuals who surf regularly (say, twenty days each year) wearing only swimsuits face a 10^{-4} increased lifetime risk of skin cancer, and that the risk increases more than linearly with additional years of surfing. You have also found that the use of T-shirts and sunscreen will reduce this risk to about 10^{-7}. How would you communicate these findings to surfers? How do you think your message would be received? Would you expected different responses at beaches where (i) none of the surfers wear T-shirts; (ii) half wear T-shirts; and (iii) most wear T-shirts?

Problem 10-4. Ranking the Risks: Are the Experts Right?

People's perceptions of risks differ along a number of demographic factors, including age, lifestyle, and education. Table 10-3 (from Fischhoff et al 1978) summarizes how four different groups of people rated the risk (to people in the U.S. as a whole) of dying from each of thirty potential hazards. The groups were 30 college students from Eugene, Oregon; 40 members of the Eugene, Oregon League of Women Voters (LOWV); 25 business and professionals from the Eugene, Oregon "Active Club"; and 15 risk experts from around the country.

Each individual was asked to rate the problems on a numerical scale, based on number of deaths expected from each potential hazard. They were asked to be exhaustive; that is, to consider all possible causes of death from each activity.

a. Are most college students and members of the LOWV aware that more people are killed by any one of motor vehicles, handguns, smoking, and motorcycles than by nuclear power each year? If so, what are some other explanations for the higher ranking of nuclear power?

b. Suggest some possible explanations for the large differences in rankings for nuclear power and x-rays.

c. Suggest some possible explanations for the large differences in rankings for contraception.

Solution 10-4

Solution 10-4a

Ignorance of the numbers is, of course, one possible (and often cited) explanation. However, many well-informed individuals are concerned about nuclear power, while many poorly informed individuals favor it. It is unlikely that there would be much disagreement over historical data on average year risks. More likely, then, the high ranking of nuclear power as a risk reflects a variety of qualitative features of "risk"

Table 10-3 Comparison of Group Rankings of Thirty Potential Hazards

	Group 1: LOWV	Group 2: college students	Group 3: Active club members	Group 4: experts
Nuclear power	1	1	8	20
Motor vehicles	2	5	3	1
Handguns	3	2	1	4
Smoking	4	3	4	2
Motorcycles	5	6	2	6
Alcoholic beverages	6	7	5	3
General (private) aviation	7	15	11	12
Police work	8	8	7	17
Pesticides	9	4	15	8
Surgery	10	11	9	5
Fire fighting	11	10	6	18
Large construction	12	14	13	13
Hunting	13	18	10	23
Spray cans	14	13	23	26
Mountain climbing	15	22	12	29
Bicycles	16	24	14	15
Commercial aviation	17	16	18	16
Electric power	18	19	19	9
Swimming	19	30	17	10
Contraceptives	20	9	22	11
Skiing	21	25	16	30
X-rays	22	17	24	7
High-school and college football	23	26	21	27
Railroads	24	23	20	19
Food preservatives	25	12	28	14
Food coloring	26	20	30	21
Power mowers	27	28	25	28
Prescription antibiotics	28	21	26	24
Home appliances	29	27	27	22
Vaccinations	30	29	29	25

Source: Lichtenstein et al. 1978.

including high uncertainty, high dread, historical misinformation, and justifiable distrust of the institutions that must manage nuclear power. Also salient are such unresolved scientific, technical, and political issues as a long-term solution to high-level radioactive waste disposal, and the difficulty of keeping nuclear power and nuclear weapons production separate. At the same time, it is possible that the experts are more familiar with nuclear waste, and that familiarity causes them to be less (perhaps inadequately) concerned.

Solution 10-4b

The League of Women Voters, college students, and Active Club Members (groups 1, 2, and 3) all rank nuclear power at or near the top of the danger list, while "experts" rank it near the bottom of the list. Thus, despite the obvious connections from physics, "nuclear power" and "X-rays" evidently mean very different things to the general public and to experts. To most people, X-ray exposure is associated with dental visits. Even if dental work is not considered pleasant, it is seen as commonplace, and X-rays are not seen as a significant risk.

Solution 10-4c

Risk of death from contraceptive use is interesting in that the responses are distinctly clustered, with college students and experts in close agreement, while members of the League of Women Voters and Active Club members are also in agreement, but at a significantly lower overall ranking. Less clear, however, are the reasons for the clustering. College students, at an age where contraceptive choices are often critically important, are likely to be influenced by individual horror stories which often circulate on college campuses. The high ranking given to contraceptives by experts is likely a result of data in the risk literature and therefore a far less visceral or personal reaction than that of students.

On the other hand, the LOWV and Active Club members are likely to be of similar ages, educational backgrounds, and socioeconomic standings. The coordination between these two groups may result from sampling individuals who hold common knowledge and experience about contraceptives.

Problem 10-H. Implications of Differing Perspectives

What are some possible implications of table 10-3 for policy making and risk communication?

Problem 10-I. Risk, Trust, and Rationality

As discussed in solution 10-4a, many highly trained technical people believe that the health risks associated with nuclear power are quite low. They conclude that public concerns, which grew significantly in the 1970s, are irrational. However, a 1965 report commissioned by the Atomic Energy Commission "found a worst-case maximum of 45,000 prompt fatalities, and property damage many times larger than the [original 1957] WASH-740 found, but again without accompanying probability estimates. The AEC suppressed public access to this study 'to avoid great difficulties in obtaining public acceptance of nuclear energy' (Mulvihill et al. 1965)" (Hohenemser et al. 1992). The reasoning was that placing a credible numerical value on the worst-case scenario was impossible, even though that probability was thought to be extremely small. When the study became public in the early 1970s, the reaction was what the experts had predicted: opposition to nuclear power.

a. Discuss what this scenario suggests about public rationality in the case of nuclear power.

b. The experts who drafted the 1965 report believed that the real risks from nuclear power were very small, at least as compared to the risks from other energy sources. Did the experts in this case define "real risk" in the same way as the public? Discuss how these definitions might differ, and how this might impact the policy arena.

c. In 1970, the Freedom of Information Act allowed citizens to access nonclassified public documents (as well as many classified documents). It was through a Freedom of Information request by the Union of Concerned Scientists that the 1965 report became public (Hohenemser et al. 1992). What impact might the suppression of information have had on public perception of nuclear power? Could the AEC have handled the 1965 report differently?

Problem 10-J. Who Is More Concerned?

a. Flynn et al. (1994) find that white males consider a large number of activities, including cigarette smoking, blood transfusions, and radon in homes, to be less risky than do white women and males of other racial backgrounds. Discuss the impact that this finding might have in the policy-making arena.

b. Now consider that the perception by white males is bimodal, with a relatively small number of white males holding "low-risk" views that are sufficiently powerful to move the mean for white males as a group. Again, discuss the impact that this finding might have in the policy-making arena.

Problem 10-5. The Availability Bias

A number of biases and heuristics (decision rules) affect the way people perceive risks. The "availability bias" causes us to think that one risk is worse than another if we can more readily identify it. One situation in which this affects decisions on a large scale is when relatively rare events, such as airplane accidents or "flesh-eating" bacteria, are given more media coverage than are more common events, such as automobile accidents and old-age stroke deaths. The bias is minimized when people are aware that it may exist, and as a consequence adjust for it.

In many cases, the biases discussed here and in the rest of the chapter are based on questions asked of students in experimental situations. However, the results are testable hypotheses, and like any of the other methods in this book, they can be illuminated by well designed studies and additional data.

There are about 4 million deaths in the United States each year. Estimate the number of deaths that are caused each year by each of the following:

- All cancer
- Botulism
- Diabetes
- Pregnancy
- All disease

Solution 10-5

The actual numbers (for 1978, but they remain at the same order of magnitude) are

- All cancer   300,000
- Botulism   3
- Diabetes   80,000
- Pregnancy   800
- All disease   1,200,000

In 1978, Lichtenstein et al. asked a large number of people to provide estimates for these and other causes of death, and compared them to actual numbers of deaths.[1] They found that people consistently underestimated larger causes of death and overestimated small causes. Figure 10-2 recreates Lichtenstein et al.'s comparison of actual and estimated numbers of death from forty-one causes, including the six presented here.

The discrepancy between actual recorded number of deaths and estimated number of deaths may be partly explained by the "availability heuristic" (Tversky and Kahneman 1973), which is the tendency to base the expected likelihood of an event on the ability to recall instances of that event. Consequently, when relatively frequent events receive less media attention than do relatively rare events, the former are undervalued and the latter are magnified. Particularly salient on this chart are botulism and diabetes; while estimates tend to be within one order of magnitude, the actual numbers are almost four orders of magnitude

[1]Note that the "actual number of deaths" is not necessarily exact, but is based on imperfect national statistics (Fischhoff 1998).

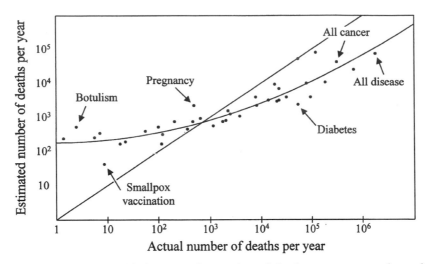

Figure 10-2. Relationship between the number of deaths per year as estimated by a large survey of lay people and the actual number of deaths as taken from national statistics (points). The straight line indicates what this relationship would be if there were a perfect match between estimated and actual numbers of deaths. After Lichtenstein et al. 1978.

apart. The estimates more closely reflect the frequency with which the two come up in public discourse than the numbers of deaths they cause. Also worth noting is that none of the listed causes was estimated to be below ten[2] nor greater than ten thousand. The public perception of the number of fatalities from a cause that is both rare and widely discussed in the press—such as ebola—is likely to be even more inflated from the true value.

Problem 10-K. Aggregation

The problem is developed from Slovic et al. 1979. Determine which of the following two options you would expect people to find safer, and explain why.

---

[2] It is possible that individual estimates were below ten, since the points represent aggregate estimates. However, a 95% confidence interval around botulism does not include ten. Also, the study used geometric means to keep rare high estimates from precluding a low mean.

a. A safety system at a refinery described as having 15 potential failure points, each with an independent failure probability of 10^{-5}

b. A safety system at a refinery described as having an overall failure probability of 15×10^{-5}

Problem 10-6. How Intuitive Are Statistics? The Case of Electromagnetic Fields and Cancer

Electromagnetic fields (EMFs) are generated by the motion of charged particles and are themselves a result and manifestation of light, or photons. Today, man-made EMFs are everywhere. Low-level EMFs (typically less than 300 Hz) surround many everyday appliances, such as microwaves and cellular phones, as well as electrical wiring and above-ground high-voltage wiring. Electromagnetic fields have been the subject of numerous toxicological and epidemiological studies (over one hundred), and no mechanisms have yet been discovered that would initiate carcinogenesis. The initial epidemiological studies that suggested a link between EMF exposure and cancer are themselves complicated because it is not the observed magnetic fields but electrical wiring codes entering homes and apartments that showed a weak correlation with cancer.

In fact, the fields that have been suggested as causing cancers, including childhood leukemia, are lower than the Earth's magnetic field, to which we are constantly exposed. At a theoretical level, there is as yet no known causal mechanism, since the energy associated with such fields is not above the background "noise." New mechanisms are proposed all the time, such as the possible effects of EMFs on melatonin production and the possibility that transient fields that exist briefly when appliances are first turned on trigger an as yet unknown effect.

This problem looks at some epidemiological studies, and how they have been interpreted to suggest a link between EMFs and cancer.

Electromagnetic field exposure can be estimated in a number of ways. One approach is to measure current in the wiring of a home or workplace and make some assumptions about the relationship between that and actual exposures. Another is to measure magnetic field strength (in units of teslas).

Figures 10-3 and 10-4 show the "odds ratio," (equivalent to relative risk) for childhood leukemia as a function of magnetic field and household wiring configuration for 638 children with childhood leukemia and 620 matched controls (from Linet et al. 1997).

  a. How would you interpret these findings? How might you expect others to interpret them, and why?

   "Meta-analysis" involves combining a large number of studies on a single issue to see whether, in aggregate, they contain information that no single study provides alone. Ideally, data from many studies would be pooled into a larger data base, but often methodological issues (such as measurements and end points) make pooling impossible. Figure 10-5 contains the means and 95% statistical confidence intervals for the relative risk of leukemia from twenty-seven studies of residential exposure to elevated EMFs based on a range of measurement estimates (Loh et al. 1997).

  b. Given twenty independent studies, in how many would you expect to find effects at a 95% confidence level?

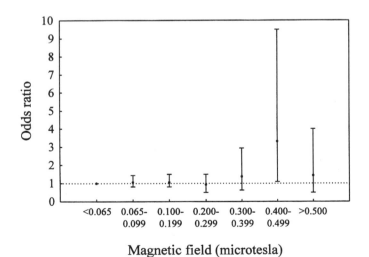

Figure 10-3. Odds ratio for childhood leukemia as a function of household magnetic field level (after Linet et al. 1997). An odds ratio of one means no effect. The bars show 95% statistical confidence intervals and the dots are means.

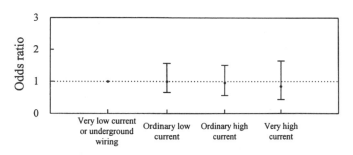

Household wiring configuration

Figure 10-4. Odds ratio for childhood leukemia as a function of household wiring configuration (after Linet et al. 1997). The bars show 95% statistical confidence intervals and the dots are means.

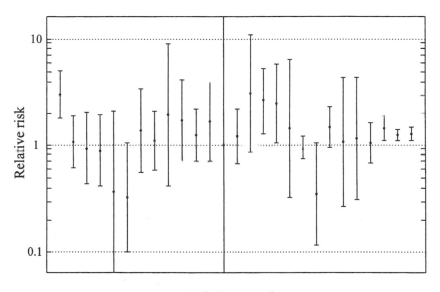

Individual studies

Figure 10-5. Relative risks of leukemia due to elevated household EMF exposures, from twenty-seven independent studies. An RR of one means no effect. The bars show 95% statistical confidence intervals and the dots are geometric means.

c. In light of your answer to (b), how do you interpret figure 10-5? How might others interpret these data?

Solution 10-6

Solution 10-6a

This group of studies provides compelling evidence that the association between childhood leukemia and household EMFs, if it exists, is too weak to be detected. As discussed in chapter 6, one should always be skeptical about low relative risks for qualitative reasons, and in this case the argument is well buttressed by quantitative measures.

Solution 10-6b

To get a feel for the degree to which statistical tests can be both helpful and deceptive, think in broad terms about this question. A 95% confidence interval means that 5% of the time the test will be inaccurate. Out of twenty tests, you would therefore expect one to show a false positive result, and should not be surprised if several do. For perspective, a baseball umpire who is right 95% of the time will call one ball in every twenty a strike. Note, however, that twenty "independent" studies might be independent in the sense of having different samples, yet share similar methodologies, which can skew them all in one direction.

Solution 10-6c

The above set of studies suggest that one cannot infer a correlation between household EMFs and childhood leukemia. Nonetheless, it is not uncommon to see the few studies that do suggest a link cited and the others ignored.

However, to argue that a causal link between EMFs and childhood leukemia does not appear to have a scientific basis does not mean that those who are concerned about EMFs are necessarily "antiscience" or "irrational." Rather, a deeper understanding of how people make decisions, distinguishing between beliefs (which may be erroneous) and values (which cannot be "erroneous," even if we disagree with them) will provide a better starting point for deciding how to influence those decisions. Risk communication using properly selected and presented information can change peoples' beliefs, although no amount of information is likely to influence values and attitudes. Importantly, ignoring beliefs and values, and providing technically correct information inappropriately, can lead to results contrary to those intended.

Problem 10-L. Alternative Interpretations

How would you respond to someone who interprets figure 10-5 as suggesting that elevated EMFs cause leukemia? As suggesting a need for further research?

Problem 10-7. Use of Point Estimates versus Distributions

The tension between the need to make decisions despite uncertainty and the need to fully characterize a risk, including quantitative and qualitative uncertainty, pervades risk decision making. Camerer and Kunreuther (1989) argue that "introducing confidence intervals into the policy making process to deal with uncertainty may be pointless[3] in many situations unless the policy analyst eliminates the point estimate itself. A specific number has a vividness and simplicity which make it an inevitable focus of public debate." This question explores the implications of the Camerer and Kunreuther conundrum.

Suppose that you have calculated the potential risk associated with each of five different approaches to reducing urban smog, $V, W, X, Y,$ and Z. Assume that costs for the five options are approximately the same. Doing nothing is also an option, but it too has costs and risks. Figure 10-6 depicts some information on the expected risk reduction for each option.

a. Which option do you expect would be picked given each of the following decision criteria:

i. "Maximin," or minimize the worst possible outcome

---

[3] Kunreuther assures us that the pun was not intended, at least not by him!

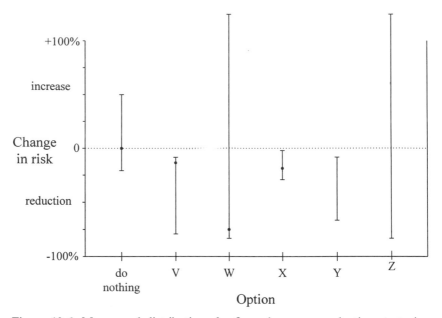

Figure 10-6. Means and distributions for five urban smog reduction strategies (*V–Z*), plus the "do nothing" option. The bars represent 95% confidence intervals, and the dots represent expected outcomes. Distributions only are provided for options (*Y*) and (*Z*).

    ii. Lowest expected value

    iii. Most certain outcome

    iv. No prior decision rule

  b. Would you expect decision makers to interpret options *V* and *Y* differently? Why, and how might this affect decisions?

Solution 10-7a(i)

The maximin decision would require one to go with either option *V* or option *Y*, since these two have the lowest maximum possible values. This decision criterion does not differentiate between option *V* and option *Y*. Decision makers might perceive the differences in several ways, including

- that option Y is the better choice because the expected value is likely to be lower than the expected value of option V (if, for example, one assumes a normal or uniform distribution), or
- that option V is the better option because it has the lowest minimum value.

Solution 10-7a(ii)

Option W is the preferred option if expected value is the only criterion

Solution 10-7a(iii)

Option X is the most certain outcome in the sense that it has the tightest distribution.

Solution 10-7a(iv)

There are a number of possible choices if a decision rule has not been predetermined, and arguments can be made for most of the options. Any of the above arguments could apply. Option W is the most likely to be unacceptable because of the possibly extreme (high) outcome. A risk-seeking individual might prefer option Z because it could have the lowest expected value (although the opposite is true as well).

Solution 10-7b

The main difference between option V and option Y is that we know a bit more about option V. It is possible that decision makers will assume that the expected value of option Y is lower than that for option V, even though there is no information to support assumptions about expected values. On the other hand, the "certainty" of the midpoint estimate could inspire greater faith in option V, since uncertainty information may be perceived as more honest but less competent (Johnson and Slovic 1995).

Perceptions of trust and credibility are essential elements for a message to be accepted (e.g., Renn and Levine 1991). Paul Slovic (1993) asserts that trust "is fragile. It is typically created rather slowly, but can be destroyed in an instant . . . lost trust may never be regained." Individuals and groups often have good reasons for deep distrust, and are likely to pick and choose from available information when trying to convince others of their decisions. Similarly, "genuine dilemmas" arise when information sources and interests are the same (Leiss 1995).

Trust in both private industry and government has declined in the past several decades (Slovic 1993). At the same time, faith in science and technology has changed but has not necessarily diminished (Renn and Levine 1991). Since much risk information comes from these two sources, the findings about trust imply that how risk information is processed by the public is likely to be significantly influenced by limited and contingent trust.

Nonetheless, there are some general lessons about how to improve perception of trustworthiness (in addition to acting in a trustworthy fashion!). Peters et al. (1997) have found that perceptions of knowledge/expertise, openness/honesty, and concern/care are the keys to trust and credibility. They also find that different groups need to focus on different aspects: industry needs to exhibit concern and care, government commitment, and citizen groups knowledge and expertise. However, they further warn that perceptions are far from invariant across the population.

Less intuitive factors are also important. While many risk experts believe that they are on the strongest footing when presenting quantitative information, people are more willing to listen when experts are open to discussing qualitative aspects of risk as well (Renn 1991). This is consistent with the more general concept known to psychologists as "reciprocation"; people are more willing to compromise when they perceive compromise on the part of others (Cialdini 1993). Explicit mention of motives is also influential (Renn and Levine 1991): recipients of the message may perceive not discussing motive as deliberate and deceptive, and they may also assign some other, perhaps more nefarious, motive.

Problem 10-M. Alternative Options

Repeat problem 10-7b above for options W and Z. Why might preferences differ from those between V and Y?

Problem 10-N. When Is the New Leaf Turned?

A company with a long history of environmental violations and records falsification establishes a new position of vice-president of Environmental Affairs. She forms a new staff including a public relations officer and a community environmental liaison, and establishes a twenty-four-hour hotline to process community complaints. The company issues a press release explaining how this represents a "new era" in environmental management.

   a. To what extent do you think these actions will affect public perception of the company?
   b. Compare what you would expect the overall effects to be from the following three scenarios for one year later:

   i. The number of violations has not changed, and a new lawsuit is filed by the EPA for past falsified records. The public relations officer spends his time writing press releases, and the community liaison's main contact is with city and other government officials, as well as formal organization lunches and dinners. Calls to the hotline go to an answering service, and are rarely followed up.
   ii. The number of violations has not changed, and a new lawsuit is filed by the EPA for past falsified records. The public relations officer creates a flier describing the facility's operations and environmental challenges and organizes an open house at one of the major facilities. The community liaison responds to hotline calls in a timely fashion (often in person) and organizes monthly community meetings, which other company officials attend.
   iii. The number of violations decreases significantly, and the company quickly accepts responsibility for its actions and settles a

lawsuit filed by the U.S. EPA for past falsified records. The public relations officer creates a flier describing the facility's operations and environmental challenges and organizes an open house at one of the major facilities. The community liaison responds to hotline calls in a timely fashion (often in person) and organizes monthly community meetings, which other company officials attend.

Problem 10-8. What Will They Think It Means?

A significant challenge for the U.S. FDA is to figure out how to communicate to a diverse population the benefits and risks of both prescription and nonprescription medications. Currently, most medicines are sold with long descriptions in minuscule fonts. Many people never look at this information, suggesting that more information may actually be less. At the same time, a substantial number of people using medicines are illiterate or cannot read English.

a. One major at-risk group are pregnant women. An icon like the one in figure 10-7 was proposed as a way to demonstrate that a particular medicine should not be taken by pregnant women. Discuss possible problems with this icon.

b. List some goals of medicine labeling. Suggest some alternatives, and discuss their strengths and weaknesses.

Figure 10-7. Proposed "not for use by pregnant women" icon for problem 10-8.

Solution 10-8

Solution 10-8a

In some ways, this seems to be a compelling approach. The figure is clearly a pregnant woman, and the "circle with an X" symbol is widely understood, and is standard in the pictograms provided by U.S. Pharmacopeia (U.S. Pharmacopeia 1998). This message would be much more likely to reach a diverse population, especially the illiterate and non-English speakers. In fact, even the highly educated might be more likely to receive this message than a tiny warning hidden among many others.

However, a single major objection could remove this icon as an option: it could easily be interpreted as "birth control" In fact, if one had not first heard the intended purpose, the reinterpretation might come even more readily. And if one were looking for birth control rather than thinking of teratogenic effects, the likelihood of misinterpretation would probably be even greater. What one expects the content of a message to be can have a strong effect on what message is received.

This leads to an interesting anecdote from a group trying to find a way to keep people away from high-level radioactive waste depositories for the next ten thousand years. When one participant suggested the use of threatening warning signs and symbols, another noted how effective this strategy was in *attracting* people to the Egyptian pyramids (Benford 1994).

Solution 10-8b

Why are medicine labels so complicated? There are several explanations. The overarching purpose is the idea that consumers should be fully informed so that they can make good decisions. Part of this is for consumer benefit, but in addition, manufacturers need to be able to show that the consumer was informed to avoid lawsuits.

Problem 10-O. What To Tell Them?

This problem was contributed by Baruch Fischhoff. Using the procedures for risk analysis, how would you choose the information to place on a label and the priorities to give such information? How would you balance people's need to know what really matters 'and the company's need to demonstrate that they have revealed everything?

_____

Problem 10-P. Alternative Labeling

Suggest a label that would clearly and efficiently indicate that pregnant women should not use a medication without consulting a physician.

_____

Problem 10-Q. Situational Differences

Suppose you were tasked with appointing a ten-person committee to establish a new risk management approach for the situations listed below. Who would you pick to be on the committee? Where would you expect to get an analysis, and how would you choose what analysis is needed? How could you increase the likelihood that the decisions generated by this group would be accepted by the relevant publics and decision makers?

a. For all ecological risks in the United States
b. For epidemic diseases in developing countries
c. For motor vehicle risks in a small mid-west town
d. For personal safety at a major university, located in an urban setting
e. For personal safety at a major university, located in an isolated setting
f. For worker health and safety in a medium-sized paint manufacturing facility in Nairobi, Kenya

_____

Problem 10-9. Saccharin and Alar: Why the Difference?

By the mid-1970s a number of laboratory animal toxicological tests had been performed on both saccharin (an artificial sweetener with a long history of use in popular consumer products) and Alar (the brand name for daminozide, a pesticide used primarily on pear and apple crops). The two substances had similar toxicity profiles, including the following:

- Laboratory tests indicated that both are carcinogenic to animals at high levels
- Extrapolations such as those developed in chapter 5 indicated that there were similar but very small (10^{-5} to 10^{-7}) risks at "normal" consumption levels
- Methodological problems with each set of tests cast some suspicion on validity
- The responsible government agencies (FDA in the case of saccharin and the EPA in the case of Alar) concluded that studies indicated low-dose human carcinogenesis, and regulated accordingly
- Saccharin has a TD_{50} = 2143 mg/kg (rats), and a can of diet soda contains 95 mg of saccharin. The FDA has extrapolated this to suggest a 3×10^{-8} risk per can of diet soda, or a 4.8×10^{-4} lifetime risk for someone drinking a can each day for forty years.
- EPA found a lifetime risk to the general public of about 5×10^{-5} from a variety of food products in average diets in the 1980s (note that Alar has since been banned). The Natural Resources Defense Council (Sewell et al. 1989) made public statements that children, who tend to eat more of the types of foods that contain Alar residues, faced about a 3×10^{-4} lifetime cancer risk.

In the late 1970s, the FDA decided to ban saccharin under the Delaney clause of the Food, Drug and Cosmetics act. The Delaney clause (overturned in 1997) banned the use of any food additive found to be a potential human carcinogen. Considerable public outcry convinced Congress to pass specific legislation exempting saccharin.

In the mid 1980s, after a number of hearings, the EPA concluded that the amount of Alar reaching consumers was too small to warrant banning the pesticide. Several years later, the Natural Resources Defense Council (NRDC) presented analysis indicating that, because they weigh less and eat much more apple sauce and juice, children are exposed to much higher concentrations of Alar than the average person. At the same time, children are often more susceptible to environmental insults than are adults. Public outcry following the release of this study, accompanied by a public perception of foot-dragging at the EPA, eventually convinced Uniroyal (the company that made Alar) to withdraw it from the market.

    a. Given the similarities in carcinogenic potential and dose levels for the two chemicals, suggest some differences that would explain the public support for saccharin and outcry against Alar.

    b. One of the popular arguments against banning saccharin was that a person would have to drink gallons of diet soda every day to consume a dose equivalent to the animals in the study. Discuss this perception in the context of what you have learned about toxicological studies.

Solution 10-9

Solution 10-9a

Several differences may come into play here. The saccharin case is anomalous—usually a "cancer" risk elicits a strong negative reaction regardless of numerical value. Possible explanations include that people were accustomed to using saccharin (familiar risks are often perceived as less important than the unfamiliar), saccharin is considered a voluntary risk (people can opt for nondiet sodas), and diabetics considered the lack of sugar to be an offsetting benefit.

In contrast, Alar was seen as an involuntary risk (individuals could only avoid it by avoiding apple products altogether), and a risk to a sensitive subgroup (children). In addition, the approach to communications taken by both apple growers and the EPA contributed to distrust: the growers made claims of "Alar free" apples when tests showed this not to be true, and the EPA persisted in arguing that Alar was not a significant risk, without effectively communicating how and why they arrived at this conclusion.

Solution 10-9b

As we have seen, the problems with extrapolation from high to low dose are many, but this statement does not address those problems. Rather, it ridicules the decision to make such extrapolations without addressing the reasons behind the extrapolation, which are to identify possible individual effects at low doses and population effects due to aggregate doses.

Problem 10-R. When To Spin?

a. Discuss why high-to-low dose extrapolation was subject to public ridicule in the case of saccharin while in the case of Alar, the high-to-low dose extrapolations were accepted?

b. Why didn't Uniroyal or the apple growers make public comments about how much applesauce a person would have to eat to get doses identical to the laboratory animals? Could such arguments have been effective in promoting public support for Alar? Why or why not?

Problem 10-10. Can or Should "Zero Risk" Be a Goal?

While minimizing risks is an entirely appropriate goal in most decision contexts, it is impossible to reduce all risks to zero. One challenge to effective risk reduction is that what may appear to be risk elimination from one perspective is a marginal reduction from another perspective. This problem takes a close look at two "risk elimination" issues.

a. Consider the following two options, and suggest how much of a price reduction would entice you

  i. to switch from a household disinfectant with a 4×10^{-7} risk of causing a fatal condition if used regularly to one with a 5×10^{-7} risk?

  ii. to switch from a household disinfectant with no known risk of causing a fatal condition if used regularly to one with a 1×10^{-7} risk?

b. Should the government ban all household products that are known to have the potential to kill?

Solution 10-10

Solution 10-10a

The exact values that most people would give to (i) and (ii) are not as important as the observation that most people would put a finite value on (i) but would find (ii) unacceptable at any markdown (provided that the options are presented independently). In one respect, this seems reasonable, since zero risk is an attractive goal. However, since this is only one of many carcinogen hazards that an individual faces, the marginal change is a more compelling decision criterion. Again, framing is a dominant element in this decision process.

Solution 10-10b

How the government can and should deal with potentially injurious products is not a simple issue. We conclude this book with such a question because by now, it should be clear that there is no clearly "right" answer. Part of the answer lies in the question of marginal changes in both aggregate and individual risks. Part lies in what the overall, measurable costs and benefits may be. Part lies in the alternatives—are we increasing unknown risks by banning known risk? Yet another part lies in distributional issues, and whether and under what conditions it is reasonable for one group to impose risks on another.

Wrapping Up: Putting the Pieces Together

We began this book with a simple definition of risk analysis: a purposeful way to organize information, including an explicit analysis of what we don't know. Such a broad definition suggests the use of a set of problems to ground the discussion of risk in real applications. We have selected a set of problems that illustrate the methodological, analytical, and topical areas where the tools of risk analysis not only answer specific questions, but more importantly build a framework in which problems from diverse areas of health, technological, engineering, and environmental analysis can be evaluated and compared.

At the same time, risk analysis is by no means limited to the applications we have chosen as examples. We could have just as easily devoted the particulars of the problems presented and solved in this book to financial, investment, and business applications, or agricultural and meteorological forecasting. Another whole set of interesting applications and problems lurks in these areas (MeVay et al. 1997), and may find their way into later editions of this book. Our interests and therefore our examples focus on the environment and environmental problem solving; the methods we illustrate are not, however, restricted to this domain.

Should We Risk It? approaches risk analysis as a set of flexible tools that can be used to understand both the human and natural environment. In working through the problems and topics in this book—dose-response analysis, the risks associated with concentrations of various substances, statistical methods, uncertainty, Bayes' theorem and Monte Carlo methods, toxicological extrapolations, epidemiology, individual and group exposure assessment, and risk communication, perception, and management—we have focused attention within each problem onto a specific method or tool. This "building block" or, as we called it in the introduction, "walk before you run," approach seems particularly appropriate to a book intended both as a course text and also as a research methods tool. In the ten chapters of *Should We Risk It?* we applied an expanding set of methods to an ever-expanding universe of applications. We have also attempted to demonstrate that much of the analysis that often appears obscure in policy discussions is in fact relatively transparent and analytically straightforward.

This does not imply that risk analysis is always a transparent process. In the complexity of real world situations, imperfect or wholly absent data, tensions between scientific findings and political forces, and competing ideologies often overshadow (but also enrich and enliven!) the analytic aspects. How analysis fits into real decisions—or, as many in the U.S. Environmental Protection Agency are fond of saying, "where the rubber meets the road,"—reflects the crux of the analysis that we have tried to present in *Should We Risk It?* The problems we selected and worked are largely based on real-world problems and actual, and often imperfect, data. The later chapters begin increasingly to integrate the methods into the broader and less well-defined decision-making process. That said, the real "rubber meets the road" test is the degree to which you find the book useful not only in classroom exercises and applications, but as a reference volume you consult as you encounter problems where these tools can be applied. In defining the role of a "citizen scientist" or "policy scientist," Frank von Hippel once commented that "if you find an important problem worth your time, you should be able to develop the expertise to become the expert." *Should We Risk It?* will be a success if it can be used to help such citizen scientists develop one set of tools to examine and evaluate important environmental risks.

The importance, and excitement, of maintaining that level of real world contact in the problems in *Should We Risk It?* is a task that we would like to shamelessly share with you, the reader. We want to encourage a dialogue about what types of problems you have found most useful in courses, and public and private sector employment in environmental, health, and technological policy analysis and policy-making positions. There is a *Should We Risk It?* web page (http://socrates.berkeley.edu/erg/swri) where additional information on the problems, alternate solutions, and new, unsolved problems can be found. Most importantly, we hope that your experience in using this book and finding new applications, as well as your own needs for new tools, encourages you to suggest—or better yet to fully write—problems that can be shared with the risk analysis community.

References

Benford, G., (1994). "A Scientist's Notebook" *Fantasy and Science Fiction* (February)

Camerer, C. F. and Kunreuther, H. (1989) "Decision processes for low probability events: Policy implications" *Journal Policy of Analysis and Management* 8(4):565–592.

Cialdini, R. B. (1993). *Influence: Science and Practice*. Third Ed. New York, NY: Addison-Wesley Publishers.

Fischhoff, B. (1991). "Value elicitation: Is there anything in there?" *American Psychologist*, 46:835–47.

———. (1998). Personal communication.

———. (1995). "Risk perception unplugged: Twenty years of process." *Risk Analysis* 15(2):137–145.

Fischhoff, B., Watson, S., and Hope, C. (1984). "Defining risk." *Policy Sciences* 17:123–39.

Flynn, J., Kasperson, R., Kunreuther, H., and Slovic, P. (1989). "Time to rethink nuclear waste storage," *Issues in Science and Technology* 8(4):42–48.

Flynn, J., Slovic, P., and Mertz, C. (1994). "Gender, race and perception of environmental health risks." *Risk Analysis* 14(6):1101–08.

Hohenemser, C., Goble, R., and Slovic, P. (1992). "Nuclear power." In Hollander, J. M. (Ed.) *The Energy-Environment Connection*. Covello, CA: Island Press.

Johnson, B. and Slovic, P. (1995). "Presenting uncertainty in health risk assessment: Initial studies of its effects on risk perception and trust." *Risk Analysis* 15(4):485–94.

Kahneman, D., and Tversky, A. (1979). Prospect theory: An analysis of decision making under risk. *Econometrica* 47:263–91.

Kunreuther, H. and Slovic, P. forthcoming. "Risk, media and stigma."

Leiss, W. (1995). "'Down and dirty': the use and abuse of public trust in risk communication." *Risk Analysis* 15(6):685–92.

Lichtenstein, S., Slovic P., Fischhoff, B., Layman, M., and Combs, B. (1978). "Judged frequency of lethal events." *Journal of Experimental Psychology: Human Learning and Memory*, 4:551–78.

Linet, M. S., Hatch, E., Kleinerman, R., Robison, L., Kaune, W., Friedman, D., Severson, R., Haines, C., Hartsock, C., Niwa, S., Wacholder, S., and Tarone, R. (1997). "Residential exposure to magnetic fields and acute lymphoblastic leukemia in children." *New England Journal of Medicine* 337(1):1–7.

Loh, Y., Shlyakhter, A., and Wilson, R. (1997). "Electromagnetic fields and the risk of leukemia and brain cancer: a summary of epidemiological literature." *Technology: Journal of the Franklin Institute* 224A:3–21.

MeVay, J., Makarov, V., and Turner, C. (1997). *Understanding Value-at-Risk Methodologies*. New York, NY: Chase Financial Services.

Mulvihill, R. J., Arnold, D. R., Bloomquist, C. E., and Epstein, B. (1965). "Analysis of United States power reactor accident probability." PRC R-695. Los Angeles: Planning Research Corporation. (Unpublished draft from the file "WASH 740 update," Public Documents Room, Nuclear Regulatory Commission, Washington.) As referenced in Hohenemser et al. 1992.

National Research Council. (1996). *Understanding Risk*. Washington, DC: National Academy Press.

Otway, H. (1992), "Public wisdom, expert fallibility." In Krimsky, S. and Golding, D. (Eds.) *Social Theories of Risk*. New York, NY: Praeger.

Peters, R., Covello, V., and McCallum, D. (1997). "The determinants of trust and credibility in environmental risk communication: an empirical study," *Risk Analysis* 17(1):43–54.

Renn, O. (1991). "Strategies of risk communication: Observations from two participatory experiments." In Kasperson, R. E. and Stallen, P. J. M. (Eds.) *Communicating Risks To The Public* Boston, MA: Kluwer Academic Publishers.

Renn, O. and Levine, D. (1991). "Credibility and trust in risk communication." In Kasperson, R. E. and Stallen, P. J. M. (Eds.) *Communicating Risks To The Public*. Boston, MA: Kluwer Academic Publishers.

Sewell, B. H., Whyatt, R. M., Hathaway, J., and Mott, L. (1989). *Intolerable Risk: Pesticides in our Children's Food*. New York, NY: Natural Resources Defense Council.

Slovic, P. (1987). "Perception of Risk." *Science* 236:279–85.

———. (1993). "Risk perception and trust." In Molak, V. (Ed.) *Fundamentals of Risk Analysis and Risk Management*. New York, NY: CRC Lewis Publishers.

Slovic, P., Lichtenstein, S., and Fischhoff, B. (1979). "Rating the risks," *Environment* 21(3):14–20, 36–39.

Tversky, A. and Kahneman, D. (1973). "Availability: a heuristic for judging frequency and probability," *Cognitive Psychology* 4:342–55.

U.S. Pharmacopeia. (1998). http://www.usp.org/did/pgrams/describe.htm (accessed May 16, 1998).

Vaughn, E. (1995). "The significance of socioeconomic and ethnic diversity for the risk communication process." *Risk Analysis* 15(2):169–80.

Walker, V. R. "Direct inference, probability, and a conceptual gulf in risk communication." *Risk Analysis* 15(5):603–9.

Wolfe, B. (1996). "Why environmentalists should support nuclear power." *Issues in Science and Technology* 12(4):55–60.

Appendix A

Z-Scores

Table A-1 Cumulative Area under the Standard Normal Distribution

	0	1	2	3	4	5	6	7	8	9
−3.0	0.0013	0.0013	0.0013	0.0012	0.0012	0.0011	0.0011	0.0011	0.0010	0.0010
−2.9	0.0019	0.0018	0.0018	0.0017	0.0016	0.0016	0.0015	0.0015	0.0014	0.0014
−2.8	0.0026	0.0025	0.0024	0.0023	0.0023	0.0022	0.0021	0.0021	0.0020	0.0019
−2.7	0.0035	0.0034	0.0033	0.0032	0.0031	0.0030	0.0029	0.0028	0.0027	0.0026
−2.6	0.0047	0.0045	0.0044	0.0043	0.0041	0.0040	0.0039	0.0038	0.0037	0.0036
−2.5	0.0062	0.0060	0.0059	0.0057	0.0055	0.0054	0.0052	0.0051	0.0049	0.0048
−2.4	0.0082	0.0080	0.0078	0.0075	0.0073	0.0071	0.0069	0.0068	0.0066	0.0064
−2.3	0.0107	0.0104	0.0102	0.0099	0.0096	0.0094	0.0091	0.0089	0.0087	0.0084
−2.2	0.0139	0.0136	0.0132	0.0129	0.0125	0.0122	0.0119	0.0116	0.0113	0.0110
−2.1	0.0179	0.0174	0.0170	0.0166	0.0162	0.0158	0.0154	0.0150	0.0146	0.0143
−2.0	0.0228	0.0222	0.0217	0.0212	0.0207	0.0202	0.0197	0.0192	0.0188	0.0183
−1.9	0.0287	0.0281	0.0274	0.0268	0.0262	0.0256	0.0250	0.0244	0.0239	0.0233
−1.8	0.0359	0.0351	0.0344	0.0336	0.0329	0.0322	0.0314	0.0307	0.0301	0.0294
−1.7	0.0446	0.0436	0.0427	0.0418	0.0409	0.0401	0.0392	0.0384	0.0375	0.0367
−1.6	0.0548	0.0537	0.0526	0.0516	0.0505	0.0495	0.0485	0.0475	0.0465	0.0455
−1.5	0.0668	0.0655	0.0643	0.0630	0.0618	0.0606	0.0594	0.0582	0.0571	0.0559
−1.4	0.0808	0.0793	0.0778	0.0764	0.0749	0.0735	0.0721	0.0708	0.0694	0.0681
−1.3	0.0968	0.0951	0.0934	0.0918	0.0901	0.0885	0.0869	0.0853	0.0838	0.0823
−1.2	0.1151	0.1131	0.1112	0.1093	0.1075	0.1056	0.1038	0.1020	0.1003	0.0985
−1.1	0.1357	0.1335	0.1314	0.1292	0.1271	0.1251	0.1230	0.1210	0.1190	0.1170
−1.0	0.1587	0.1562	0.1539	0.1515	0.1492	0.1469	0.1446	0.1423	0.1401	0.1379
−0.9	0.1841	0.1814	0.1788	0.1762	0.1736	0.1711	0.1685	0.1660	0.1635	0.1611
−0.8	0.2119	0.2090	0.2061	0.2033	0.2005	0.1977	0.1949	0.1922	0.1894	0.1867
−0.7	0.2420	0.2389	0.2358	0.2327	0.2296	0.2266	0.2236	0.2206	0.2177	0.2148
−0.6	0.2743	0.2709	0.2676	0.2643	0.2611	0.2578	0.2546	0.2514	0.2483	0.2451
−0.5	0.3085	0.3050	0.3015	0.2981	0.2946	0.2912	0.2877	0.2843	0.2810	0.2776
−0.4	0.3446	0.3409	0.3372	0.3336	0.3300	0.3264	0.3228	0.3192	0.3156	0.3121
−0.3	0.3821	0.3783	0.3745	0.3707	0.3669	0.3632	0.3594	0.3557	0.3520	0.3483
−0.2	0.4207	0.4168	0.4129	0.4090	0.4052	0.4013	0.3974	0.3936	0.3897	0.3859
−0.1	0.4602	0.4562	0.4522	0.4483	0.4443	0.4404	0.4364	0.4325	0.4286	0.4247
0.0	0.5000	0.4960	0.4920	0.4880	0.4840	0.4801	0.4761	0.4721	0.4681	0.4641
0.0	0.5000	0.5040	0.5080	0.5120	0.5160	0.5199	0.5239	0.5279	0.5319	0.5359
0.1	0.5398	0.5438	0.5478	0.5517	0.5557	0.5596	0.5636	0.5675	0.5714	0.5753
0.2	0.5793	0.5832	0.5871	0.5910	0.5948	0.5987	0.6026	0.6064	0.6103	0.6141
0.3	0.6179	0.6217	0.6255	0.6293	0.6331	0.6368	0.6406	0.6443	0.6480	0.6517
0.4	0.6554	0.6591	0.6628	0.6664	0.6700	0.6736	0.6772	0.6808	0.6844	0.6879
0.5	0.6915	0.6950	0.6985	0.7019	0.7054	0.7088	0.7123	0.7157	0.7190	0.7224

Table A-1 (*Continued*)

	0	1	2	3	4	5	6	7	8	9
0.6	0.7257	0.7291	0.7324	0.7357	0.7389	0.7422	0.7454	0.7486	0.7517	0.7549
0.7	0.7580	0.7611	0.7642	0.7673	0.7704	0.7734	0.7764	0.7794	0.7823	0.7852
0.8	0.7881	0.7910	0.7939	0.7967	0.7995	0.8023	0.8051	0.8078	0.8106	0.8133
0.9	0.8159	0.8186	0.8212	0.8238	0.8264	0.8289	0.8315	0.8340	0.8365	0.8389
1.0	0.8413	0.8438	0.8461	0.8485	0.8508	0.8531	0.8554	0.8577	0.8599	0.8621
1.1	0.8643	0.8665	0.8686	0.8708	0.8729	0.8749	0.8770	0.8790	0.8810	0.8830
1.2	0.8849	0.8869	0.8888	0.8907	0.8925	0.8944	0.8962	0.8980	0.8997	0.9015
1.3	0.9032	0.9049	0.9066	0.9082	0.9099	0.9115	0.9131	0.9147	0.9162	0.9177
1.4	0.9192	0.9207	0.9222	0.9236	0.9251	0.9265	0.9279	0.9292	0.9306	0.9319
1.5	0.9332	0.9345	0.9357	0.9370	0.9382	0.9394	0.9406	0.9418	0.9429	0.9441
1.6	0.9452	0.9463	0.9474	0.9484	0.9495	0.9505	0.9515	0.9525	0.9535	0.9545
1.7	0.9554	0.9564	0.9573	0.9582	0.9591	0.9599	0.9608	0.9616	0.9625	0.9633
1.8	0.9641	0.9649	0.9656	0.9664	0.9671	0.9678	0.9686	0.9693	0.9699	0.9706
1.9	0.9713	0.9719	0.9726	0.9732	0.9738	0.9744	0.0750	0.9756	0.9761	0.9767
2.0	0.9772	0.9778	0.9783	0.9788	0.9793	0.9798	0.9803	0.9808	0.9812	0.9817
2.1	0.9821	0.9826	0.9830	0.9834	0.9838	0.9842	0.9846	0.9850	0.9854	0.9857
2.2	0.9861	0.9864	0.9868	0.9871	0.9875	0.9878	0.9881	0.9884	0.9887	0.9890
2.3	0.9893	0.9896	0.9898	0.9901	0.9904	0.9906	0.9909	0.9911	0.9913	0.9916
2.4	0.9918	0.9920	0.9922	0.9925	0.9927	0.9929	0.9931	0.9932	0.9934	0.9936
2.5	0.9938	0.9940	0.9941	0.9943	0.9945	0.9946	0.9948	0.9949	0.9951	0.9952
2.6	0.9953	0.9955	0.9956	0.9957	0.9959	0.9960	0.9961	0.9962	0.9963	0.9964
2.7	0.9965	0.9966	0.9967	0.9968	0.9969	0.9970	0.9971	0.9972	0.9973	0.9974
2.8	0.9974	0.9975	0.9976	0.9977	0.9977	0.9978	0.9979	0.9979	0.9980	0.9981
2.9	0.9981	0.9982	0.9982	0.9983	0.9984	0.9984	0.9985	0.9985	0.9986	0.9986
3.0	0.9987	0.9987	0.9987	0.9988	0.9988	0.9989	0.9989	0.9989	0.9990	0.9990

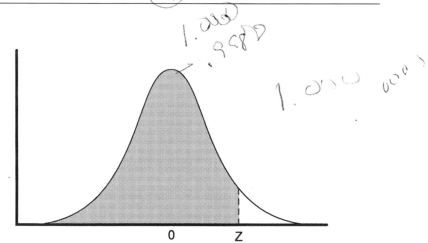

Figure A-1. Z-scores.

Appendix B

Student's *t*-Test

Table B-1 Upper Percentiles for Student's *t*-Distribution with *n* Degrees of Freedom

n	0.2	0.15	0.1	0.05	0.025	0.01	0.005
1	1.3760	1.9630	3.0780	6.3138	12.7060	31.821	63.6570
2	1.0610	1.3860	1.8860	2.9200	4.3027	6.965	9.9248
3	0.9780	1.2500	1.6380	2.3534	3.1825	4.541	5.8409
4	0.9410	1.1900	1.5330	2.1318	2.7764	3.747	4.6041
5	0.9200	1.1560	1.4760	2.0150	2.5706	3.365	4.0321
6	0.9060	1.1340	1.4400	1.9432	2.4469	3.143	3.7074
7	0.8960	1.1190	1.4150	1.8946	2.3646	2.998	3.4995
8	0.8890	1.1080	1.3970	1.8595	2.3060	2.896	3.3554
9	0.8830	1.1000	1.3830	1.8331	2.2622	2.821	3.2498
10	0.8790	1.0930	1.3720	1.8125	2.2281	2.764	3.1693
11	0.8760	1.0880	1.3630	1.7959	2.2010	2.718	3.1058
12	0.8730	1.0830	1.3560	1.7823	2.1788	2.681	3.0545
13	0.8700	1.0790	1.3500	1.7709	2.1604	2.650	3.0123
14	0.8680	1.0760	1.3450	1.7613	2.1448	2.624	2.9768
15	0.8660	1.0740	1.3410	1.7530	2.1315	2.602	2.9467
16	0.8650	1.0710	1.3370	1.7459	2.1199	2.583	2.9208
17	0.8630	1.0690	1.3330	1.7396	2.1098	2.567	2.8982
18	0.8620	1.0670	1.3300	1.7341	2.1009	2.552	2.8784
19	0.8610	1.0660	1.3280	1.7291	2.0930	2.539	2.8609
20	0.8600	1.0640	1.3250	1.7247	2.0860	2.528	2.8453
21	0.8590	1.0630	1.3230	1.7207	2.0796	2.518	2.8314
22	0.8580	1.0610	1.3210	1.7171	2.0739	2.508	2.8188
23	0.8580	1.0600	1.3190	1.7139	2.0687	2.500	2.8073
24	0.8570	1.0590	1.3180	1.7109	2.0639	2.492	2.7969
25	0.8560	1.0580	1.3160	1.7081	2.0595	2.485	2.7874
26	0.8560	1.0580	1.3150	1.7056	2.0555	2.479	2.7787
27	0.8550	1.0570	1.3140	1.7033	2.0518	2.473	2.7707
28	0.8550	1.0560	1.3130	1.7011	2.0484	2.467	2.7633
29	0.8540	1.0550	1.3110	1.6991	2.0452	2.462	2.7564
30	0.8540	1.0550	1.3100	1.6973	2.0423	2.457	2.7500
31	0.8535	1.0541	1.3095	1.6955	2.0395	2.453	2.7441
32	0.8531	1.0536	1.3086	1.6939	2.0370	2.449	2.7385
33	0.8527	1.0531	1.3078	1.6924	2.0345	2.445	2.7333
34	0.8524	1.0526	1.3070	1.6909	2.0323	2.441	2.7284
35	0.8521	1.0521	1.3062	1.6896	2.0301	2.438	2.7239

Table B-1 (*Continued*)

n	0.2	0.15	0.1	0.05	0.025	0.01	0.005
36	0.8518	1.0516	1.3055	1.6883	2.0281	2.434	2.7195
37	0.8515	1.0512	1.3049	1.6871	2.0262	2.431	2.7155
38	0.8512	1.0508	1.3042	1.6860	2.0244	2.428	2.7116
39	0.8510	1.0504	1.3037	1.6849	2.0227	2.426	2.7079
40	0.8507	1.0501	1.3031	1.6839	2.0211	2.423	2.7045
41	0.8505	1.0498	1.3026	1.6829	2.0196	2.421	2.7012
42	0.8503	1.0494	1.3020	1.6820	2.0181	2.418	2.6981
43	0.8501	1.0491	1.3016	1.6811	2.0167	2.416	2.6952
44	0.8499	1.0488	1.3011	1.6802	2.0154	2.414	2.6923
45	0.8497	1.0485	1.3007	1.6794	2.0141	2.412	2.6896
46	0.8495	1.0483	1.3002	1.6787	2.0129	2.410	2.6870
47	0.8494	1.0480	1.2998	1.6779	2.0118	2.408	2.6846
48	0.8492	1.0478	1.2994	1.6772	2.0106	2.406	2.6822
49	0.8490	1.0476	1.2991	1.6766	2.0096	2.405	2.6800
50	0.8489	1.0473	1.2987	1.6759	2.0086	2.403	2.6778
51	0.8448	1.0471	1.2984	1.6753	2.0077	2.402	2.6758
52	0.8486	1.0469	1.2981	1.6747	2.0067	2.400	2.6738
53	0.8485	1.0467	1.2978	1.6742	2.0058	2.399	2.6718
54	0.8484	1.0465	1.2875	1.6736	2.0049	2.397	2.6700
55	0.8483	1.0463	1.2972	1.6731	2.0041	2.396	2.6683
56	0.8481	1.0461	1.2969	1.6725	2.0033	2.395	2.6666
57	0.8480	1.0460	1.2967	1.6721	2.0025	2.393	2.6650
58	0.8479	1.0458	1.2964	1.6716	2.0017	2.392	2.6633
59	0.8478	1.0457	1.2962	1.6712	2.0010	2.391	2.6618
60	0.8477	1.0455	1.2959	1.6707	2.0003	2.390	2.6603
61	0.8476	1.0454	1.2957	1.6703	1.9997	2.389	2.6590
62	0.8475	1.0452	1.2954	1.6698	1.9990	2.388	2.6576
63	0.8474	1.0451	1.2952	1.6694	1.9984	2.387	2.6563
64	0.8473	1.0449	1.2950	1.6690	1.9977	2.386	2.6549
65	0.8472	1.0448	1.2948	1.6687	1.9972	2.385	2.6537
66	0.8471	1.0447	1.2945	1.6683	1.9966	2.384	2.6525
67	0.8471	1.0446	1.2944	1.6680	1.9961	2.383	2.6513
68	0.8470	1.0444	1.2942	1.6676	1.9955	2.382	2.6501
69	0.8469	1.0443	1.2940	1.6673	1.9950	2.381	2.6491
70	0.8468	1.0442	1.2938	1.6669	1.9945	2.381	2.6480
71	0.8468	1.0441	1.2936	1.6666	1.9940	2.380	2.6470
72	0.8467	1.0440	1.2934	1.6661	1.9935	2.379	2.6459
73	0.8466	1.0439	1.2933	1.6660	1.9931	2.378	2.6450
74	0.8465	1.0438	1.2931	1.6657	1.9926	2.378	2.6640
75	0.8465	1.0437	1.2930	1.6655	1.9922	2.377	2.6431
76	0.8464	1.0436	1.2928	1.6652	1.9917	2.376	2.6421
77	0.8464	1.0435	1.2927	1.6649	1.9913	2.376	2.6413
78	0.8463	1.0434	1.2925	1.6646	1.9909	2.375	2.6406
79	0.8463	1.0433	1.2924	1.6644	1.9905	2.374	2.6396
80	0.8462	1.0432	1.2922	1.6641	1.9901	2.374	2.6388
81	0.8461	1.0431	1.2921	1.6639	1.9897	2.373	2.6380
82	0.8460	1.0430	1.2920	1.6637	1.9893	2.372	2.6372

Table B-1 (*Continued*)

n	0.2	0.15	0.1	0.05	0.025	0.01	0.005
83	0.8460	1.0430	1.2919	1.6635	1.9890	2.372	2.6365
84	0.8459	1.0429	1.2917	1.6632	1.9886	2.371	2.6357
85	0.8459	1.0428	1.2916	1.6630	1.9883	2.371	2.6350
86	0.8458	1.0427	1.2915	1.6628	1.9880	2.370	2.6343
87	0.8458	1.0427	1.2914	1.6626	1.9877	2.370	2.6336
88	0.8457	1.0426	1.2913	1.6624	1.9873	2.369	2.6329
89	0.8457	1.0426	1.2912	1.6622	1.9870	2.369	2.6323
90	0.8457	1.0425	1.2910	1.6620	1.9867	2.368	2.6316
91	0.8457	1.0424	1.2909	1.6618	1.9864	2.368	2.6310
92	0.8456	1.0423	1.2908	1.6616	1.9861	2.367	2.6303
93	0.8456	1.0423	1.2907	1.6614	1.9859	2.367	2.6298
94	0.8455	1.0422	1.2906	1.6612	1.9856	2.366	2.6292
95	0.8455	1.0422	1.2905	1.6611	1.9853	2.366	2.6286
96	0.8454	1.0421	1.2904	1.6609	1.9850	2.366	2.6280
97	0.8454	1.0421	1.2904	1.6608	1.9848	2.365	2.6275
98	0.8453	1.0420	1.2903	1.6606	1.9845	2.365	2.6270
99	0.8453	1.0419	1.2902	1.6604	1.9843	2.364	2.6265
100	0.8452	1.0418	1.2901	1.6602	1.9840	2.364	2.6260
∞	0.8400	1.0400	1.2800	1.6400	1.9600	2.330	2.5800

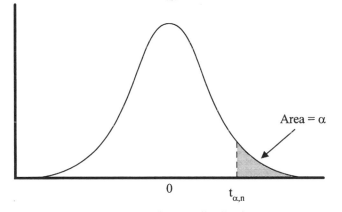

Figure B-1. Student's *t*-distribution.

Appendix C

Chi-Squared Distribution

Table C-1 Upper and Lower Percentiles of Chi-Squared Distributions (with k Degrees of Freedom)

k	0.01	0.025	0.05	0.1	0.9	0.95	0.975	0.99
1	0.00016	0.00098	0.00393	0.0158	2.706	3.841	5.024	6.635
2	0.0201	0.0506	0.103	0.211	4.605	5.991	7.378	9.210
3	0.115	0.216	0.352	0.584	6.251	7.815	9.348	11.345
4	0.297	0.484	0.711	1.064	7.779	9.488	11.143	13.277
5	0.554	0.831	1.145	1.610	9.236	11.070	12.832	15.086
6	0.872	1.237	1.635	2.204	10.645	12.592	14.449	16.812
7	1.239	1.690	2.167	2.833	12.017	14.067	16.013	18.475
8	1.646	2.180	2.733	3.490	13.362	15.507	17.535	20.090
9	2.088	2.700	3.325	4.168	14.684	16.919	19.023	21.666
10	2.558	3.247	3.940	4.865	15.987	18.307	20.483	23.209
11	3.053	3.816	4.575	5.578	17.275	19.675	21.920	24.725
12	3.571	4.404	5.226	6.304	18.549	21.026	23.336	26.217
13	4.107	5.009	5.892	7.042	19.812	22.362	24.736	27.688
14	4.660	5.629	6.571	7.790	21.064	23.685	26.119	29.141
15	5.229	6.262	7.261	8.547	22.307	24.996	27.488	30.578
16	5.812	6.908	7.962	9.312	23.542	26.296	28.845	32.000
17	6.408	7.564	8.672	10.085	24.769	27.587	30.191	33.409
18	7.015	8.231	9.390	10.865	25.989	28.869	31.526	34.805
19	7.633	8.907	10.117	11.651	27.204	30.144	32.582	36.191
20	8.260	9.591	10.851	12.443	28.412	31.410	34.170	37.566
21	8.897	10.283	11.591	13.240	29.615	32.671	35.479	38.932
22	9.542	10.982	12.338	14.041	30.813	33.924	36.781	40.289
23	10.196	11.688	13.091	14.848	32.007	35.172	38.076	41.638
24	10.586	12.401	13.848	15.659	33.196	36.415	39.364	42.980
25	11.524	13.120	14.611	16.473	34.382	37.652	40.646	44.314
26	12.198	13.844	15.379	17.292	35.563	38.885	41.923	45.642
27	12.879	14.573	16.151	18.114	36.741	40.113	43.194	46.963
28	13.565	15.308	16.928	18.939	37.916	41.337	44.461	48.278
29	14.256	16.047	17.708	19.768	39.087	42.557	45.722	49.588
30	14.953	16.791	18.493	20.599	40.256	43.773	46.979	50.892
31	15.655	17.539	19.281	21.434	41.422	44.985	48.232	52.191
32	16.362	18.291	20.072	22.271	42.585	46.194	49.480	53.486
33	17.073	19.047	20.867	23.110	43.745	47.400	50.725	54.776
34	17.789	19.806	21.664	23.952	44.903	48.602	51.966	56.061
35	18.509	20.569	22.465	24.797	46.059	49.802	53.203	57.342

Table C-1 (*Continued*)

k	0.01	0.025	0.05	0.1	0.9	0.95	0.975	0.99
36	19.233	21.336	23.269	25.643	47.212	50.998	54.437	58.619
37	19.960	22.106	24.075	26.492	48.363	52.192	55.668	59.892
38	20.691	22.878	24.884	27.343	49.513	53.384	56.985	61.162
39	21.426	23.654	25.695	28.196	50.660	54.572	58.120	62.428
40	22.164	24.433	26.509	29.051	51.805	55.758	59.342	63.691
41	22.906	25.215	27.326	29.907	52.949	56.942	60.561	64.950
42	23.650	25.999	28.144	30.765	54.090	58.124	61.777	66.206
43	24.398	26.785	28.965	31.625	55.230	59.304	62.990	67.459
44	25.148	27.575	29.787	32.487	56.369	60.481	64.201	68.709
45	25.901	28.366	30.612	33.350	57.505	61.656	65.410	69.957
46	26.657	29.160	31.439	34.215	58.641	62.830	66.617	71.201
47	27.416	29.956	32.268	35.081	59.774	64.001	67.821	72.443
48	28.177	30.755	33.098	35.949	60.907	65.171	69.023	73.683
49	28.941	31.555	33.930	36.818	62.038	66.339	70.222	74.919
50	29.707	32.357	34.764	37.689	63.167	67.505	71.420	76.154

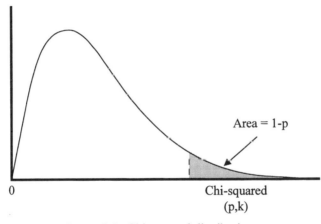

Figure C-1. Chi-squared distributions.

Index